An Introduction to Graphene and Carbon Nanotubes

An Introduction to Graphene and Carbon Nanotubes

John Edward Proctor, Daniel Alfonso Melendrez Armada, and Aravind Vijayaraghavan

CRC Press
Taylor & Francis Group
Boca Raton London New York

CRC Press is an imprint of the
Taylor & Francis Group, an **informa** business

CRC Press
Taylor & Francis Group
6000 Broken Sound Parkway NW, Suite 300
Boca Raton, FL 33487-2742

First issued in paperback 2020

© 2017 by Taylor & Francis Group, LLC
CRC Press is an imprint of Taylor & Francis Group, an Informa business

No claim to original U.S. Government works

ISBN 13: 978-0-367-57390-4 (pbk)
ISBN 13: 978-1-4987-5179-7 (hbk)

Library of Congress Cataloging-in-Publication Data

Names: Proctor, John Edward, 1981 or 1982- author. | Melendrez Armada, Daniel Alfonso, author. | Vijayaraghavan, Aravind, 1979- author.
Title: An introduction to graphene and carbon nanotubes / John Edward Proctor, Daniel Alfonso Melendrez Armada, and Aravind Vijayaraghavan.
Description: Boca Raton, FL : CRC Press, Taylor & Francis Group, [2016] | Includes bibliographical references and index.
Identifiers: LCCN 2016033579| ISBN 9781498751797 (hardback ; alk. paper) | ISBN 1498751792 (hardback ; alk. paper) | ISBN 9781498751810 (e-book) | ISBN 1498751814 (e-book)
Subjects: LCSH: Graphene. | Carbon nanotubes. | Nanostructured materials.
Classification: LCC TA455.G65 P76 2016 | DDC 620.1/15--dc23
LC record available at https://lccn.loc.gov/2016033579

Visit the Taylor & Francis Web site at
http://www.taylorandfrancis.com

and the CRC Press Web site at
http://www.crcpress.com

Contents

Preface

This textbook has been conceived as a result of developing and teaching a third-year undergraduate lecture course on graphene and carbon nanotubes at the University of Salford, and as such the text draws on this experience. It has been written such that substantial parts of the text should be accessible to a third-year undergraduate who has a working knowledge of basic quantum mechanics topics such as the solutions to the Schrödinger equation for the hydrogenic atom but who is embarking on the study of solid-state physics for the first time. Topics in the text are then developed to a suitable level for the final-year undergraduate and the student who is embarking on postgraduate research.

Single-walled carbon nanotubes and graphene have been the subject of intense scientific research since their discovery in 1993 and 2004, respectively. A large amount of literature is available on these materials, and one may ask why another textbook is justified. It is therefore appropriate to outline what this textbook offers.

In the case of graphene, a number of texts are available giving good treatments of the fundamental electronic properties of graphene at a level suitable for a postgraduate student, the massless nature of its charge carriers and their description using the Dirac equation. On the other hand, many topics in graphene research are less well covered in the literature. A research student who would like to read an honest assessment of the advantages and disadvantages of different methods to synthesize or exfoliate graphene may struggle to find this.

In the case of single-walled carbon nanotubes, there have been major developments in the past five years which are not covered in the traditional textbooks on this subject. In particular, the synthesis of nanotubes with a specific (n,m) assignment has become possible. It is therefore now possible to synthesize a sample of nanotubes which all have identical optical and electronic properties. This arguably removes the biggest roadblock to commercial applications of nanotubes in electronics. In addition, our assignment of (n,m) indices to nanotubes using optical methods has become far more accurate due to improvements in our understanding of the role of excitons in nanotubes and the effect of the tube's environment on both the exciton energy levels and the phonon energies.

For both nanotubes and graphene, existing texts are written at a level suitable for postgraduate students and the potential to use these systems as excellent simple models to teach undergraduate students basic concepts in solid-state physics and physical chemistry remains untapped. This book seeks to address this using nanotubes and graphene as models to teach concepts such as molecular orbital theory, tight binding theory and the Laue treatment of diffraction.

Last but not least, I would like to draw the reader's attention to the bibliography. Separate from the extensive reference list, the bibliography is a list of around 60 original research papers, review papers and textbooks which I have selected as being worthwhile for further reading. Approaching a field such as graphene or carbon nanotubes can be daunting for the student at the start of his or her research career, who is still developing the skills of searching the scientific literature. At the time of writing, typing "graphene" into Google Scholar produces 646,000 results and the student new to the field may struggle to see the wood for the trees. The bibliography attempts to assist the student by providing a list of a small number of high-quality and readable publications which contain useful discussion relating to the underlying science, instead of merely presenting new results. The bibliography does not attempt to be an exhaustive list of every good publication on these topics. In addition, it would not be possible to prepare a text such as this without drawing upon the standard textbooks in solid state physics and materials chemistry; these are also cited in the bibliography.

I thank my co-authors Daniel Melendrez Armada and Dr. Aravind Vijayaraghavan. Daniel has produced most of the figures for this text; I have been fortunate to work with someone who is capable of turning my disorganized sketches and instructions into high-quality figures that bring the subject to life. Whilst the bulk of this work focusses on the basic physics and chemistry of graphene and carbon nanotubes, the field is one of the most exciting research areas in modern science and the final chapter of the book, which Aravind has contributed, really does justice to this. I am confident it will inspire students to take up study and research in the field. I am also thankful to Aravind for providing critical comment on the entirety of the text.

I thank my colleagues at the University of Salford for tolerating my neglect of certain other tasks whilst writing this textbook. In addition, I thank the many friends, colleagues and other researchers in the field who have assisted in the preparation of this text through critically reviewing various chapters, providing samples or original data for many of the figures and engaging in scientific discussion on this topic: Prof. Hyeonsik Cheong, Prof. Trevor Cox, Nicholas Dominelli Whiteley, Prof. David Dunstan, Dr. Michael Finegan, Dr. John Gallop, Prof. Matthew Halsall, Sam Harrison, Jack Holguin, Prof. Marco Kralj, Jake Taylor Jones, Prof. Ado Jorio, Prof. Vladimir Kuznetsov, Dr. Márcia Lucchese,

Prof. Jossano Marcuzzo, Addison Marshall, Prof. Jannik Meyer, Prof. Ian Morrison, Prof. Nicolas Mounet, Dr. Chris Muryn, Sam Neild, Prof. Konstantin Novoselov, Geoff Parr, Dr. Andrei Sapelkin, Dr. Dean Smith, Dr. Lee Webster, Simon Wickham, Emily Woodroofe, Dr. Heather Yates, Dr. Duhee Yoon and Dr. Recep Zan.

Whilst extensive proofreading has taken place, it is perhaps inevitable that some errors remain and these are entirely my own responsibility. I would be grateful if these could be reported for correction in any future editions. Please send error reports to j.e.proctor@salford.ac.uk. Raman spectra (where not otherwise credited) are my own data, collected on a conventional single-grating Raman spectrometer (1200 or 1800 lines per inch grating) in the backscattering geometry with 532 nm laser excitation.

John E. Proctor
Bolton, United Kingdom

Useful Equations

Graphene lattice vectors:

$$\boldsymbol{a}_1 = \begin{pmatrix} \dfrac{\sqrt{3}a}{2} \\ \dfrac{a}{2} \end{pmatrix}; \quad \boldsymbol{a}_2 = \begin{pmatrix} \dfrac{\sqrt{3}a}{2} \\ -\dfrac{a}{2} \end{pmatrix} \quad \text{where } |\boldsymbol{a}_1| = |\boldsymbol{a}_2| = a = \sqrt{3}a_0$$

The accepted values for the bond length a_0 and lattice constant a are $a_0 = 1.42\,\text{Å}$ and $a = 2.46\,\text{Å}$.

Graphene reciprocal lattice vectors:

$$\boldsymbol{b}_1 = \begin{pmatrix} \dfrac{2\pi}{\sqrt{3}a} \\ \dfrac{2\pi}{a} \end{pmatrix}; \quad \boldsymbol{b}_2 = \begin{pmatrix} \dfrac{2\pi}{\sqrt{3}a} \\ -\dfrac{2\pi}{a} \end{pmatrix}$$

Graphene electronic dispersion relation:

$$E(\boldsymbol{k}) = \pm \frac{tw(\boldsymbol{k})}{1 \pm sw(\boldsymbol{k})}$$

where

$$w(\boldsymbol{k}) = \sqrt{1 + 4\cos\frac{\sqrt{3}k_x a}{2}\cos\frac{k_y a}{2} + 4\left(\cos\frac{k_y a}{2}\right)^2}$$

$$t = -3.033\,\text{eV}$$

$$s = 0.129$$

SWCNT chiral vector and diameter:

$$C_h = n\boldsymbol{a}_1 + m\boldsymbol{a}_2$$

$$d_t = \frac{|C_h|}{\pi} = \frac{a}{\pi}\sqrt{n^2 + m^2 + nm}$$

$$\cos\theta = \frac{C_h \cdot \boldsymbol{a}_1}{|C_h||\boldsymbol{a}_1|} = \frac{2n + m}{2\sqrt{n^2 + m^2 + nm}}$$

SWCNT categories:

Armchair: (n,n)
Zigzag: $(n,0)$
Chiral: (n,m) where $0 < m < n$

SWCNT translation vector:

$$T = t_1 a_1 + t_2 a_2$$

where

$$t_1 = \frac{2m+n}{p}$$

$$t_2 = -\frac{2n+m}{p}$$

p is the greatest common divisor of $(2m+n)$ and $(2n+m)$.

1 Introduction

1.1 Graphite

Graphite, the thermodynamically stable phase of carbon at ambient conditions, has been studied and used by human-kind for centuries. For instance, during the reign of Queen Elizabeth I, graphite from a large high-quality deposit near Borrowdale in the English Lake District was used as a material to line the moulds for cannonballs. This resulted in rounder and smoother cannonballs than the UK's military competitors, so production at the mine was strictly controlled by the Crown. To this day, graphite is used for an important and very diverse range of applications such as nuclear reactor moderators, pencils, electric motor brushes and addition of carbon to steel.

1.1.1 Crystal structure of graphite and graphene

In terms of atomic structure, graphite is a layered material where each layer consists of a hexagonal lattice of carbon atoms joined by strong covalent bonds. The bonds between the layers, on the other hand, are weak van der Waals bonds (Figure 1.1). A single atomic layer of graphite is called graphene. For graphene, we can define a primitive unit cell: the smallest building block from which we can construct the graphene lattice. The primitive unit cell of graphene consists of two atoms due to the hexagonal structure, which we can label as A and B (Figure 1.1). The size of the primitive unit cell of graphite depends on how the individual graphene layers stack to form the graphite crystal. Graphite is found in nature with various stacking arrangements, but in this text we will concentrate on the most common and thermodynamically stable stacking: Bernal (or ABAB) stacking. In Bernal stacking the B atom in the second layer is directly above the A atom in the first layer, and then in the third layer there is an A atom at this location, just as in the first layer. The primitive unit cell of Bernal-stacked graphite thus consists of four atoms in two adjacent layers. The graphite crystal shown in Figure 1.1 exhibits Bernal stacking.

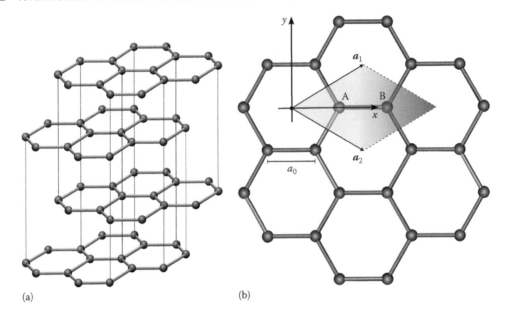

(a)   (b)

FIGURE 1.1

(a) Atomic structure of Bernal (ABAB) stacked graphite, the most common and thermo-dynamically stable form of graphite. (b) Primitive unit cell (shaded) and lattice vectors a_1 and a_2 of graphene.

We can define lattice vectors, the vectors joining equivalent points in adjacent unit cells, for mono-layer graphene. Equation 1.1 gives these (two-dimensional) lattice vectors, in terms of the sp^2 C–C bond length in graphene a_0:

$$a_1 = \begin{pmatrix} \dfrac{\sqrt{3}a}{2} \\ \dfrac{a}{2} \end{pmatrix}; \quad a_2 = \begin{pmatrix} \dfrac{\sqrt{3}a}{2} \\ -\dfrac{a}{2} \end{pmatrix} \quad \text{where } |a_1| = |a_2| = a = \sqrt{3}a_0 \quad (1.1)$$

The accepted values [1] for the bond length and lattice constant are $a_0 = 1.42\,\text{Å}$ and $a = 2.46\,\text{Å}$. We can also (Equation 1.2) define the reciprocal lattice vectors b_1 and b_2 for the graphene lattice using the relations $a_1 \cdot b_1 = 2\pi$, $a_1 \cdot b_2 = 0$, etc. [2]:

$$b_1 = \begin{pmatrix} \dfrac{2\pi}{\sqrt{3}a} \\ \dfrac{2\pi}{a} \end{pmatrix}; \quad b_2 = \begin{pmatrix} \dfrac{2\pi}{\sqrt{3}a} \\ -\dfrac{2\pi}{a} \end{pmatrix} \quad (1.2)$$

Many readers will be familiar with the concept of the recipro-cal lattice. The reciprocal lattice of a crystalline material is the Fourier transform of the real space lattice (Appendix C). Whilst the real space lattice is periodic with period determined by the lattice vectors (e.g. those defined in Equation 1.1 for graphene), the reciprocal lattice is periodic with period determined by the reciprocal lattice vectors and has units of wavevector k. Reciprocal space is also referred to as k-space. The physical significance of the reciprocal lattice is that it governs the way in which waves, and particles exhibiting wave–particle duality, propagate through the material. The relationship between energy and wavevector (the dispersion relation) for electrons and phonons propagating in the crystal is periodic with period given by the reciprocal lattice vectors, and the crystal will diffract radiation/particles when the change in wavevector upon diffraction is a reciprocal lattice vector or integer multiples thereof.

It is the role of the reciprocal lattice in determining for what scattering vectors (angles) a crystal will diffract radiation/ particles which led to the development of the reciprocal lattice concept. Readers not familiar with the reciprocal lattice concept may wish to read Section 8.1 on the Laue treatment of diffraction, which introduces the concept of the reciprocal lattice by demonstrating its role in understanding diffraction. The reciprocal lattice concept is covered in great detail in the literature [2–5].

1.1.2 Electronic properties of graphite and graphene layers

From the electronic point of view, graphite exhibits highly anisotropic behaviour. In the plane of the individual graphene layers, graphite exhibits very high conductivity, whilst in the direction normal to the graphene planes, the conductivity is somewhat lower. The electronic dispersion relation of graphite has been studied theoretically for many decades prior to the isolation of graphene [6–8]. As we shall discuss in detail in Chapters 2 through 4, graphene features covalent bonding in which three of the four valence electrons form strong directional interatomic bonds (σ-bonds) to neighbouring atoms in the graphene layer. The electrons in the σ-bonds are strongly bound into the bonds so they cannot move. The fourth valence electron, on the other hand, is responsible for weak π-bonds between neighbouring atoms in the graphene layer and weak van der Waals bonds between the graphene layers. The binding energy of these bonds is small so these electrons can easily be excited into the anti-bonding orbital (conduction band) to allow an electric current to flow.

The appropriate methodology to study the electronic dispersion relation of graphite is the tight-binding approximation, which we will use for ourselves in Chapters 3 and 4 and Appendix B. Most authors (even back to the 1940s [6]) chose to consider only a single atomic layer of graphene as a simplifying approximation and therefore assume that the electrons are confined within a single graphene layer – free to move only in two dimensions within the plane of this layer.

Using this approximation, we can obtain a result for the two-dimensional electronic dispersion relation of a single graphene layer (Equation 1.3 and Figure 1.2). We have used the results from Saito et al. [8], but similar results were obtained by Wallace in 1947 [6] and in many subsequent studies:

$$E(\mathbf{k}) = \pm \frac{t w(\mathbf{k})}{1 \pm s w(\mathbf{k})} \qquad (1.3)$$

where

$$w(\mathbf{k}) = \sqrt{1 + 4 \cos \frac{\sqrt{3} k_x a}{2} \cos \frac{k_y a}{2} + 4 \left(\cos \frac{k_y a}{2} \right)^2} \qquad (1.4)$$

$$t = -3.033 \text{ eV}$$

$$s = 0.129$$

Referring to Figure 1.2, we can see that the valence band (bonding orbitals) and conduction band (anti-bonding orbitals) meet at points in reciprocal space labelled as the K points. Hence, graphene is a zero-bandgap semiconductor: Since the conduction and valence bands meet, even an extremely small amount of thermal energy $\left(E \sim \frac{3}{2} k_B T \right)$ can excite electrons from the valence band to the conduction band. As a result,

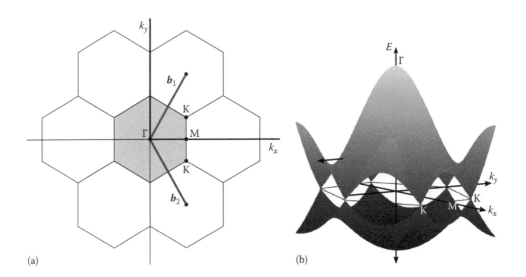

(a) (b)

FIGURE 1.2
(a) Reciprocal lattice of graphene with reciprocal lattice vectors $\mathbf{b}_1$ and $\mathbf{b}_2$ marked, first Brillouin zone shaded, high symmetry points Γ (the centre of the Brillouin zone), K and M marked. (b) Electronic dispersion relation (plot of electron energy as a function of wavevector $\mathbf{k}$ in the xy plane) for graphene with valence band (lower section) and conduction band (upper section) meeting at the K point.

there are unbound electrons in the conduction band at any temperature above absolute zero. Therefore, the calculation is in agreement with the experimental observation that graphite is an extremely good conductor within the plane of the graphene layers.

1.2   Carbon fibres

Whilst graphite has been used by humankind for many centuries, in the nineteenth century attention turned to the manufacture of graphitic carbon fibres. Carbon fibres were originally developed for use as filaments in lightbulbs (the first carbon fibre was manufactured for this purpose by Thomas Edison in 1879), but these were soon superseded by tungsten filaments and carbon fibres were instead manufactured for applications making use of their high strength. Indeed to this day, they are widely used as part of composite materials and textiles.

In terms of atomic structure, a carbon fibre (Figure 1.3) consists of a very large number of graphite flakes rolled up into a cylindrical rod. Carbon fibres have macroscopic dimensions; the curvature of the graphite flakes induced by rolling up to form the fibre is not sufficient to induce significant changes to

9 µm

FIGURE 1.3
Scanning electron micrograph of carbon fibre. (Reproduced from Marcuzzo, J.S. et al., *Mat. Res.*, 16, 137, 2013. With permission.)

the properties of the individual flakes, and a single pristine layer of graphene does not extend all the way round the fibre. This distinguishes them from multi-walled carbon nanotubes.

1.3   Buckminsterfullerene (C_{60}) and multi-walled carbon nanotubes (MWCNTs)

In 1985, a discovery was made which completely changed the face of carbon research: the synthesis of Buckminsterfullerene (C_{60}) by pulsed laser vaporization of graphite by Kroto et al. in 1985 [10]. Carbon fibres are, on an atomic scale, composed of graphite, the thermodynamically stable three-dimensional allotrope of carbon. There are three degrees of freedom for electrons and phonons. C_{60}, however, is a zero-dimensional allotrope of carbon. Electrons and phonons are confined on a microscopic level in all three dimensions. In C_{60} (Figure 1.4), a graphene lattice is formed from 60 carbon atoms in which some of the hexagons are replaced by pentagons so that the lattice forms a sphere instead of a flat sheet. Whilst C_{60} is the smallest fullerene possible (at about 0.7 nm diameter), a number of larger fullerenes have also been synthesized [11].

Following the discovery of C_{60}, it was hypothesized [12] that an extremely small carbon fibre could be manufactured in which each single pristine layer of graphene did extend all the way round the fibre and join back up on itself, and potentially the thickness of this fibre could be just a single layer of graphene. These structures are the multi-walled carbon nanotube (MWCNT; Figure 1.5) and the single-walled

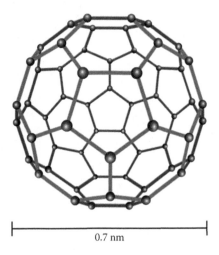

0.7 nm

FIGURE 1.4
Schematic diagram of Buckminsterfullerene, C_{60}, a zero-dimensional allotrope of carbon formed from a graphene lattice curved into a spherical shape by the replacement of some hexagons with pentagons.

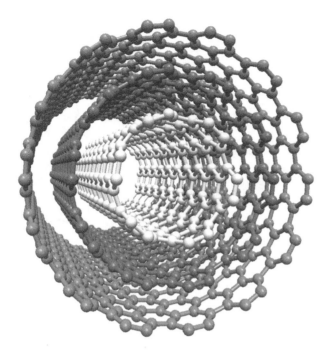

FIGURE 1.5
Schematic diagram of a multi-walled carbon nanotube, formed from continuous concentric graphene tubes. The innermost tube (highlighted in white) is often as small as 1 nm diameter.

carbon nanotube (SWCNT; Figures 1.6 through 1.8). The synthesis of MWCNTs is widely credited to Iijima in 1991 (using a laser ablation method similar to that used to synthesize fullerenes) [13], and this discovery led on to the synthesis of SWCNTs in 1993 [14,15] (Section 1.4).

It is, however, now recognized that MWCNTs were in fact synthesized many decades prior to the paper by Iijima in 1991 [13] and even to the discovery of C_{60} in 1985 [10]. The work was done by researchers in the former USSR; its significance was not recognized (the concept of "nanotechnology" did not exist in the 1950s) and indeed the work was not widely read in the West simply due to the fact that it was published in Russian. The earliest recognized synthesis of a MWCNT is from 1952 [16,17] by Radushkevich and Lukyanovich, researchers at the Institute of Physical Chemistry, USSR Academy of Science. Whilst the details have perhaps been lost forever with the passing of time, it is known that Radushkevich and Lukyanovich were working together with Dubinin on the adsorption of carbonaceous materials (e.g. materials produced by combustion of coal) on transition metals. They noticed that in some cases the structures formed were quite unusual, leading to the detailed study published in 1952 [16] unambiguously

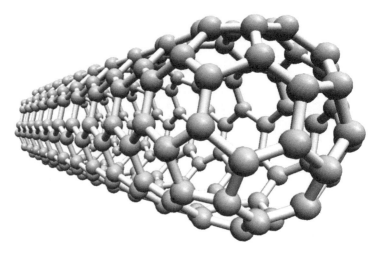

FIGURE 1.6
Schematic diagram of a (5,5) single-walled carbon nano-tube capped by half of a C_{60} molecule.

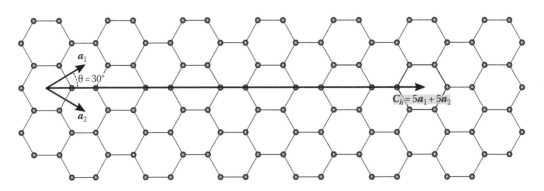

FIGURE 1.7
The graphene lattice making up a section of the (5,5) single-walled carbon nanotube with graphene lattice vectors (a_1, a_2), the chiral vector C_h and chiral angle θ marked. $C_h = 5a_1 + 5a_2$.

identifying them (using transmission electron microscopy) as what are now known as MWCNT. Transition metals are nowadays widely used as a catalyst for growth of carbon nanotubes (both multi-walled and single-walled) and transmission electron microscopy is still the most direct method of counting the number of graphitic walls making up a nanotube.

Whilst the attention of the research community swiftly moved on to SWCNT following their synthesis in 1993 [14,15], it is worthwhile to note that, as far as current commercial applications are concerned, MWCNTs are often used in preference to their single-walled counterparts due to the lower cost of production [18].

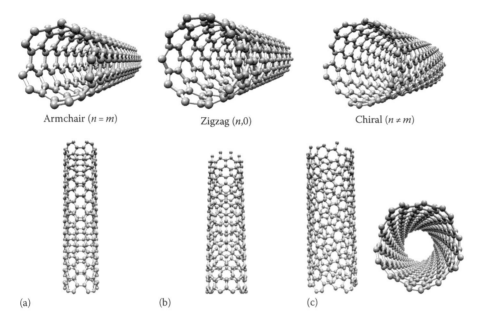

Armchair $(n = m)$ Zigzag $(n,0)$ Chiral $(n \neq m)$

(a)   (b)   (c)

FIGURE 1.8
Schematic diagrams of example achiral armchair $(n = m)$ (a), achiral zigzag $(n,0)$ (b) and chiral $(n \neq m)$ (c) single-walled carbon nanotubes (SWCNTs). The chiral SWCNT can exist in two different forms, or enantiomers, equivalent to a left- or right-handed screw, as illustrated in the view on the bottom right along the chiral SWCNT axis.

In the MWCNT, the requirement for the graphene lattice from which each layer of the tube is formed to join up on itself does have some effect on material properties – for instance, it drastically increases Young's modulus compared to that of a carbon fibre [19]. However, in the SWCNT, the continuity of the graphene lattice has a far greater effect – in particular, quantum confinement of electrons in the direction around the nanotube (the circumferential direction) is the crucial factor in determining the electronic and optical properties of the tube (Chapter 5).

1.4   Single-walled carbon nanotubes (SWCNTs)

In contrast to the lengthy and convoluted history regarding the discovery of MWCNTs, there is no controversy regarding the discovery of SWCNTs. Their discovery was made independently by two different groups of researchers (Iijima and Ichihashi at NEC Corporation in Japan, Bethune and co-workers at IBM in the United States). The findings from both groups were published simultaneously in *Nature* in 1993 [14,15].

In the SWCNT, a tube is formed from a single graphene sheet (Figure 1.6) rolled into a tube and joining back up on itself. The tube can be (and normally is) terminated by half of a fullerene molecule. The smallest SWCNT regularly synthesized is that capped by half of the smallest possible fullerene molecule, C_{60} (the (5,5) nanotube, with a diameter of 0.7 nm) [12]. Larger SWCNTs are capped by larger fullerene molecules.

We can define a vector in terms of the graphene lattice vectors to completely describe the structure of any SWCNT and predict key properties: the chiral vector. The chiral vector is a vector which circumnavigates the nanotube once, perpendicular to the tube axis. For the graphene lattice forming the tube to be continuous, the chiral vector (C_h) must consist of integer (n,m) multiples of the graphene lattice vectors a_1 and a_2 (as given in Equation 1.1). The SWCNT diameter d_t can also be obtained from the chiral vector. We can thus write

$$C_h = na_1 + ma_2 \tag{1.5}$$

$$d_t = \frac{|C_h|}{\pi} = \frac{a}{\pi}\sqrt{n^2 + m^2 + nm}$$

Figure 1.7 shows the graphene lattice forming the (5,5) SWCNT from Figure 1.6, this time in its "unrolled" form. The graphene lattice vectors and the chiral vector C_h are marked, along with the chiral angle θ. θ is the angle between the graphene lattice vector a_1 and the chiral vector, defined according to Equation 1.6. The hexagonal symmetry of the graphene lattice ensures that any SWCNT can be described with a chiral angle in the range $0 \le θ \le 30°$:

$$\cos θ = \frac{C_h \cdot a_1}{|C_h||a_1|} = \frac{2n + m}{2\sqrt{n^2 + m^2 + nm}} \tag{1.6}$$

Nomenclature has been developed (Figure 1.8) to classify SWCNTs according to the chiral angle θ and chiral indices (n,m). A SWCNT with $θ = 30°$, (n,n) is an "armchair" SWCNT. A SWCNT with $θ = 0°$, $(n,0)$ is a "zigzag" SWCNT, and a SWCNT with $0 < θ < 30°$, (n,m) where $n \ne m$ is a "chiral" SWCNT. We can see (Figure 1.8) that the names "armchair" and "zigzag" refer to the shape of the cross section of the graphene lattice circumnavigating the nanotube. The chiral term arises from the fact that SWCNTs in this category have a mirror image that is not identical to the original SWCNT. A chiral SWCNT with a given (n,m) can exist in two different forms or enantiomers, equivalent in appearance to a left-handed or right-handed screw.* The mirror image of

---

* The author once announced, as an April fools' joke, that he had secured funding to transport his chemical vapour deposition furnace from the United Kingdom to New Zealand to discover if, in New Zealand, the nanotubes would grow with the opposite helicity/"handedness" due to the Coriolis effect.

an armchair or zigzag SWCNT, on the other hand, is identical to the original tube. Hence, we describe these tubes as achiral.

The SWCNT structure can only exist on an extremely small scale. Unlike MWCNTs, in which the presence of several concentric graphene rings has a stabilizing effect, SWCNTs with a diameter greater than ~2.5 nm spontaneously collapse at ambient conditions [20,21]. This is simply because, for a larger SWCNT, the energy saving upon collapse (when the adjacent walls of the flat nanotube can form van der Waals bonds) exceeds the strain energy due to the deformation of the graphene lattice at the edges of the collapsed tube.

The (inflated) SWCNT can therefore only exist as a structure with diameter of the order of 1 nm. Hence (unlike the MWCNT), quantum confinement of electrons in the direction circumnavigating the SWCNT must always play a role in determining the electronic properties.

In contrast to the diameter of the SWCNT, which must be of the order of 1 nm, the length of the SWCNT can reach macroscopic dimensions. Typical SWCNTs have lengths in excess of 1 μm [12], and SWCNTs with lengths in excess of 10 cm [22,23] have been reported. Hence, SWCNTs have an extremely high aspect ratio (at least 10^3). We can consider them as a genuine one-dimensional allotrope of carbon, neglecting the effect of the fullerene end caps and the effect of quantum confinement of electrons in the axial direction (i.e. along the axis of the SWCNT) on their properties.

We shall see in Chapter 5 how we can predict the varying electronic properties of SWCNT by understanding them as one-dimensional objects: Due to quantum confinement of the electrons in the SWCNT, the component of the electron wavevector in the direction around the circumference of the SWCNT (the circumferential direction) can only take certain discrete values. These values are those for which the electron wave function and its derivative are continuous in the circumferential direction. In the direction along the SWCNT (the axial direction), on the other hand, the effect of quantum confinement is negligible due to the macroscopic dimension of the SWCNT in this direction. Hence, in a SWCNT, the two-dimensional electronic dispersion relation of graphene (Figure 1.2) is split into a series of one-dimensional lines through two-dimensional reciprocal space, each corresponding to an allowed value of the electron wavevector in the circumferential direction. The specific values of (n,m) dictate if one of these lines intersects with the K point, where the conduction and valence bands meet. Hence, following a detailed discussion of this topic in Chapter 5, we will predict if a given SWCNT is metallic (a zero-bandgap semiconductor) or has a finite non-zero electronic bandgap using the (n,m) assignment for the tube.

1.5 Graphene

Many (perhaps most) readers will already be aware of the landmark paper published in 2004 in the journal *Science* outlining the isolation of a single atomic layer of graphite, known as graphene [24]. The paper, published by principal authors Novoselov and Geim, described the isolation of a single atomic layer of graphite by simply peeling apart thin slices of graphite using adhesive tape until only one atomic layer is left. We will learn in Chapter 9 that the production of high-quality graphene samples for scientific research and potential applications in electronics is not quite as low-tech as it appears when described in these terms. Nonetheless, the simplicity of the process is remarkable.

Graphene completes the family of different carbon allotropes stable at ambient conditions: It is a two-dimensional allotrope of carbon. We will see in future chapters, for instance, how we understand the electronic properties of graphene by considering the electrons as particles with two degrees of freedom.

In the original 2004 paper [24], Novoselov and Geim demonstrated a simple field effect transistor fabricated from graphene. It is worthwhile to note that there was some history of synthesis of both few-layer and mono-layer graphene prior to 2004 [25–30]. However, the 2004 paper by Novoselov and Geim [24] was different to all previous works on graphene. This is because it demonstrated the synthesis/isolation of mono-layer graphene sheets of extremely high crystalline quality, in a controllable and reproducible manner, on a substrate suitable for further study, and on which the graphene crystals could be identified using an optical microscope. This is reflected in the number of citations that the publication has received (21,600 citations on Web of Science as of May 2016).

It is for these reasons that the publication sparked a gold rush of scientific interest in graphene and experiments demonstrating a whole host of remarkable properties: electrons moving at relativistic velocity, the observation of the quantum hall effect even at ambient temperature [31–33] and intrinsic strength greater than that of diamond [34]. As a result of this, Novoselov and Geim were awarded the Nobel Prize in Physics in 2010 for "ground breaking experiments regarding the two-dimensional material graphene." From the theoretical point of view, graphene is a remarkably simple system: a perfect hexagonal honeycomb lattice of carbon atoms, each with four valence electrons. In subsequent chapters, we can therefore understand in detail (using undergraduate-level physics and mathematics) all the exciting phenomena outlined earlier, and many more. We will begin in the next chapter where we will apply molecular orbital theory to understand the interatomic bonding in graphene and diamond, to understand why graphene is so strong.

References

1. DC Elias et al., *Science* **323**, 610 (2009).

2. JR Hook and HE Hall, *Solid State Physics*, Wiley-VCH, Chichester, U.K. (1998).

3. J Singleton, *Band Theory and Electronic Properties of Solids*, OUP, Oxford, U.K. (2001).

4. C Kittel, *Introduction to Solid State Physics*, Wiley-VCH, Chichester (1996).

5. NW Ashcroft and ND Mermin, *Solid State Physics*, Thomson Learning, Boston (1976).

6. PR Wallace, *Phys. Rev.* **71**, 622 (1947).

7. GS Painter and DE Ellis, *Phys. Rev. B* **1**, 4747 (1970).

8. R Saito, G Dresselhaus and MS Dresselhaus, *Phys. Rev. B* **61**, 2981 (2000).

9. JS Marcuzzo, C Otani, HA Polidoro and S Otani, *Mat. Res.* **16**, 137 (2013).

10. HW Kroto, JR Heath, SC O'Brien, RF Curl and RE Smalley, *Nature* **318**, 162 (1985).

11. PW Fowler and DE Manolopoulos, *An Atlas of Fullerenes*, Dover Publications, New York (2006).

12. R Saito, G Dresselhaus and MS Dresselhaus, *Physical Properties of Carbon Nanotubes*, ICP, London (1998).

13. S Iijima, *Nature* **354**, 56 (1991).

14. S Iijima and T Ichihashi, *Nature* **363**, 603 (1993).

15. DS Bethune et al., *Nature* **363**, 605 (1993).

16. LV Radushkevich and VM Lukyanovich, *Zurn. Fisic. Chim.* **26**, 88 (1952).

17. M Monthioux and VL Kuznetsov, *Carbon* **44**, 1621 (2006).

18. MFL De Volder, SH Tawfick, RH Baughman and AJ Hart, *Science* **339**, 535 (2013).

19. MMJ Treacy, TW Ebbesen and JM Gibson, *Nature* **381**, 678 (1996).

20. NG Chopra et al., *Nature* **377**, 135 (1995).

21. JA Elliott, JKW Sandler, AH Windle, RJ Young and MSP Shaffer, *Phys. Rev. Lett.* **92**, 095501 (2004).

22. B Peng, Y Yao and J Zhang, *J. Phys. Chem. C* **114**, 12960 (2010).

23. X Wang et al., *Nano Lett.* **9**, 3137 (2009).

24. KS Novoselov et al., *Science* **306**, 666 (2004).

25. HP Boehm, A Clauss, GO Fischer and U Hofmann, *Z. Naturforschung B* **17**, 150 (1962).

26. AJ Van Bommel, JE Crombeen and A van Tooren, *Surf. Sci.* **48**, 463 (1975).

27. I Forbeaux, J-M Themlin and J-M Debever, *Phys. Rev. B* **58**, 16396 (1998).

28. *Scientific Background on the Nobel Prize in Physics 2010*, Royal Swedish Academy of Sciences (2010).

29. ES Reich, *Nature* **468**, 486 (2010).

30. JJ Wang et al., *Appl. Phys. Lett.* **85**, 1265 (2004).

31. KS Novoselov et al., *Nature* **438**, 197 (2005).

32. Y Zhang, Y-W Tan, HL Stormer and P Kim, *Nature* **438**, 201 (2005).

33. KS Novoselov et al., *Science* **315**, 1379 (2007).

34. C Lee, X Wei, JW Kysar and J Hone, *Science* **321**, 5887 (2008).

2

Interatomic Bonding in Graphene and Diamond

In order to understand the unique, and important, mechanical and electronic properties of graphene and carbon nanotubes, it is essential to obtain a thorough knowledge of the nature of the interatomic bonding in graphene. To summarize the system in very simple terms, the C–C interatomic bond in graphene consists of bonds formed from electron clouds aligned along the axis of the bond (σ-bonds) and bonds formed from electron clouds aligned perpendicular both to the bond axis and graphene plane (π-bonds). The σ-bonds require a lot of energy to break; these bonds are responsible for the extraordinary strength of graphene (Chapter 10). The electrons in the π-bonds, on the other hand, are very weakly bound and are free to move at great velocity through the graphene layer (Chapters 3 and 4). These electrons are responsible for the extremely high electrical conductivity of graphene.

Both the σ-bonds and the π-bonds are pure covalent bonds, and their nature has been studied for decades prior to the discovery of graphene due to their wider importance in organic chemistry [1]. To understand their nature, we begin with a description of molecular orbital theory (applying it to the simplest possible molecule, the hydrogen molecule-ion). We will then apply molecular orbital theory to the bonding in graphene and diamond. The comparison between the interatomic bonding in graphene and diamond is instructive in several ways.

First, it allows us to compare the mechanical properties of graphene to those of diamond, the hardest and arguably most incompressible bulk material known. Second, when we roll the graphene lattice up into a tube to form a SWCNT or into a sphere to form a fullerene molecule, the curvature must cause the interatomic bonding to become a mixture of graphene-like and diamond-like bonding. Third, if we covalently functionalize graphene (e.g. by hydrogenation; Chapter 11), we create a material which is essentially a one-atom thick slice through the diamond lattice with (for instance) electronic properties which are broadly similar to those of diamond.

2.1   An introduction to molecular orbital theory

Molecular orbital theory describes, on a quantum mechanical level, how covalent interatomic bonds form, in what directions they form and how strong the resulting bonds are. It achieves this by allowing us to define a wave function for an electron bound to two atoms (in an interatomic bond) from the wave functions the electron would take if bound to one of the individual atoms.

Let us first apply molecular orbital theory to the simplest possible system: the hydrogen molecule-ion, consisting of two protons and one electron. We will see that it is relatively straightforward to extend this to the hydrogen molecule. To begin with, imagine the case where the two protons are a large distance apart. The electron must sit in the 1s energy level of one of the isolated protons and no bond will form. But if the protons are brought closer together, then the electron will be attracted to both protons simultaneously and finding the electronic wave functions and energy levels from first principles would be quite complex. So we simplify this task by making some approximations:

1. *The Born–Oppenheimer approximation*: We assume that the protons are fixed whilst the electron is free to move [2].

2. *The key approximation of molecular orbital theory*: That the wave function of the electron is a linear combination of the wave functions the electron would have whilst bound to either of the two isolated protons [3]. This is referred to in many texts as the LCAO (linear combination of atomic orbitals) approximation. This wave function is known as a molecular orbital, and we will hence use the notation $\Psi(MO)$.

Using these approximations, we can write the electron wave function $\Psi(MO)$ for this system as follows:

$$\Psi(MO) = \frac{1}{\sqrt{2}}\left[\Psi_A(1s) \pm \Psi_B(1s)\right] \qquad (2.1)^*$$

Here, $\Psi_A(1s)$ and $\Psi_B(1s)$ are the wave functions for an electron in the 1s energy level bound to the individual protons A and B, respectively. Note that, for symmetry reasons, the electron probability density $|\Psi(MO)|^2$ resulting from the LCAO wave function must be symmetric about a line bisecting the bond axis (marked in Figure 2.1). The wave function defined in Equation 2.1 is the only linear combination of the atomic orbitals which achieves this.

_____

* This wave function is normalized, if we assume that the overlap between $\Psi_A(1s)$ and $\Psi_B(1s)$ is small.

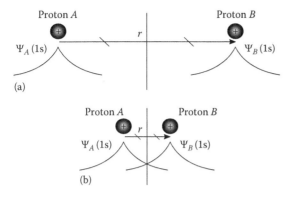

FIGURE 2.1

The hydrogen molecule-ion consists of two protons (separated by a distance r) and one electron. If r is large (a), the electron must sit in the 1s energy level bound to one of the individual protons. However, when r is reduced (b) the electron interacts with both protons leading to the formation of the hydrogen molecule-ion. In this case, we can consider the electron wave function as a linear combination of the overlapping wave functions ($\Psi_A(1s)$ and $\Psi_B(1s)$) associated with the 1s energy levels when the electron is bound to just one proton.

We now obtain an expression for the probability distribution for the electron in the $\Psi(MO)$ orbital:

$$P = \Psi\left(MO\right)^2 d\tau = \frac{1}{2}\left[\Psi_A(1s) \pm \Psi_B(1s)\right]^2 d\tau \qquad (2.2)$$

$$P = \frac{1}{2}\Psi_A\left(1s\right)^2 d\tau + \frac{1}{2}\Psi_B\left(1s\right)^2 d\tau \pm \Psi_A\left(1s\right)\Psi_B\left(1s\right)d\tau$$

We can see that if the $\Psi(MO)$ wave function takes the + sign in Equation 2.1, then according to Equation 2.2 there will be an increased probability of finding the electron in the region between the two protons, where we expect both $\Psi_A(1s)$ and $\Psi_B(1s)$ to be non-zero. We hence refer to this option as the bonding orbital:

$$\Psi\left(MO\right)_{bond} = \frac{1}{\sqrt{2}}\left[\Psi_A\left(1s\right) + \Psi_B(1s)\right] \qquad (2.3)$$

The other option is the anti-bonding orbital:

$$\Psi\left(MO\right)_{anti\text{-}bond} = \frac{1}{\sqrt{2}}\left[\Psi_A\left(1s\right) - \Psi_B(1s)\right] \qquad (2.4)$$

In this case, the probability of finding the electron equidistant between the protons drops to zero. For the example here (the hydrogen molecule-ion), it is straightforward to

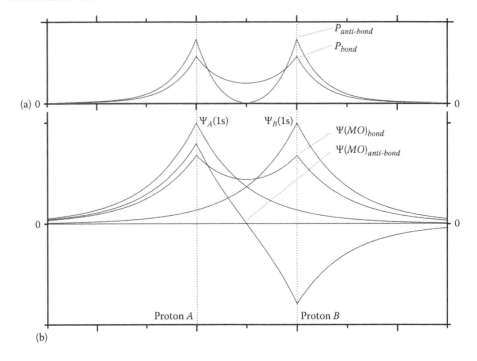

FIGURE 2.2
(a) Normalized electron probability density for the bonding and anti-bonding molecu-lar orbitals (Equation 2.2). An electron in the bonding orbital has a high probability of being found between the two protons whilst an electron in the anti-bonding orbital has a probability of being found equidistant between the two protons of exactly zero. (b) Normalized electron wave functions in the hydrogen molecule-ion system. $\Psi_A(1s)$, $\Psi_B(1s)$: wave functions for electrons bound to individual protons A and B, in the 1s energy level. $\Psi(MO)_{bond}$, $\Psi(MO)_{anti\text{-}bond}$: bonding and anti-bonding molecular orbital wave functions defined according to Equations 2.3 and 2.4, respectively.

illustrate the difference between the bonding and anti-bonding orbitals in terms of the normalized wave functions and probability of finding the electron in different regions (Figure 2.2).

Imagine that we begin with a large separation between the protons. The electron is in the 1s atomic orbital bound to one of the protons and is not interacting with the other proton. As we gradually bring the protons closer together (reduce r in Figure 2.1), the electron starts to interact with both protons so must now sit in the bonding or anti-bonding molecular orbital instead of the 1s atomic orbital. Close to a certain equilibrium separation between the protons, the energy of the system (when the electron sits in the bonding orbital) is significantly lower than it was when the separation between the protons was large and the electron was bound to just one of them. This is just as well otherwise molecules would not be stable.

On the other hand, if the electron sits in the anti-bonding orbital the energy of the system is significantly higher than it was when the separation between the protons was large and the electron was bound to just one of them. Here, we will accept these facts as intuitively clear. After all, an electron in the bonding orbital is likely to be found between the two protons so will be electrostatically attracted to both of them. An electron in the anti-bonding orbital is unlikely to be found between the two protons, maximizing the electrostatic repulsion between them. The interested reader is referred to more detailed discussions on this topic in the literature [4,5].

How can we extend this to the hydrogen (H_2) molecule? The Pauli Exclusion Principle permits the $\Psi(MO)_{bond}$ orbital to be occupied by two electrons with opposing spins, making the H_2 molecule stable. The He_2^+ molecule (two Helium nuclei and three electrons) is stable (but only just [6]) as the third electron has to sit in the anti-bonding orbital instead of the bonding orbital, at significant cost in energy. The He_2 molecule is not stable, as the cost in energy to put two electrons into the anti-bonding orbitals is too great (Figure 2.3).

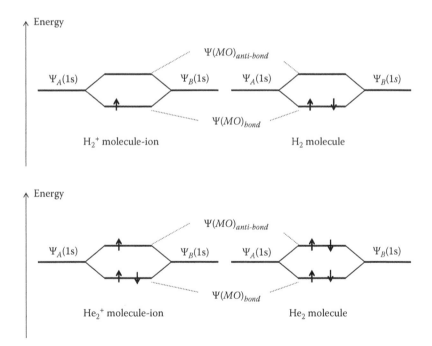

FIGURE 2.3
Occupation of bonding and anti-bonding molecular orbitals for the H_2 and He_2 molecules and molecule-ions. In the case of the He_2 molecule, the energy cost of placing two electrons in the anti-bonding orbital (as required by the Pauli exclusion principle) is too great so the molecule does not form.

2.2 Orbital hybridization in carbon and the C–C interatomic bond

The four valence electrons in the isolated carbon atom occupy the 2s orbital and the 2p orbital. The energy gap between these orbitals is very small.* The valence electrons can therefore exist partially occupying the 2s orbital and partially occupying in the 2p orbital. The wave functions of these electrons are called hybridized atomic orbitals and the phenomenon is called orbital hybridization. The hybridized atomic orbitals are formed so that the electron wave function can be the right shape to then form molecular orbitals for a desired bonding geometry, for example, the bonding in graphene or diamond.

The 2s orbital is a single wave function $\Psi(2s)$ which can be occupied by two electrons with opposing spins. The 2p orbital, however, consists of three different wave functions $\Psi(2p_x)$, $\Psi(2p_y)$, $\Psi(2p_z)$. The wave functions lead to an electron cloud oriented along the x, y and z axes, respectively.† The hybridized atomic orbitals from which the molecular orbitals form can thus be made up of different linear combinations of the $\Psi(2s)$ and $\Psi(2p_x)$, $\Psi(2p_y)$, $\Psi(2p_z)$ wave functions. The combination chosen depends on what is required to allow a molecular orbital to form with the largest overlap between the electron clouds from adjacent atoms. A larger overlap leads to more energy being saved when the bond forms, and hence a stronger bond. We refer to the three different ways in which the orbitals can mix, or hybridize, as follows:

1. sp *hybridization (e.g. carbyne or acetylene)*: Hybridized orbitals are formed from linear combinations of the $\Psi(2s)$ and $\Psi(2p_x)$ wave functions.

2. sp² *hybridization (e.g. graphene or benzene)*: Hybridized orbitals are formed from linear combinations of the $\Psi(2s)$ and $\Psi(2p_x)$, $\Psi(2p_y)$ wave functions.

3. sp³ *hybridization (e.g. diamond or methane)*: Hybridized orbitals are formed from linear combinations of the $\Psi(2s)$ and $\Psi(2p_x)$, $\Psi(2p_y)$, $\Psi(2p_z)$ wave functions.

---

* In the hydrogenic atom (i.e. Z protons and one electron), the 2s and 2p orbitals have the same energy. In a multi-electron atom such as carbon, the small energy gap between the 2s and 2p orbitals arises from the fact both sit in a potential well in which the positive charge of the nucleus is partially screened from them by the negative charge of the 1s electrons. These sit very close to the nucleus compared to the 2s and 2p electrons. An electron in the 2p orbital is likely to be found slightly further away from the nucleus than that in the 2s, therefore the screening is more effective. It is not bound quite so strongly to the nucleus as an electron in the 2s and has slightly higher energy.

† The spatial orientation or "shape" of the wave function is given by the relevant spherical harmonic function. These functions describe the angular dependence of the solutions to the Schrödinger equation for a spherically symmetric potential using spherical polar coordinates. The spherical harmonics are sketched in many reference books, for instance, *The Cambridge Handbook of Physics Formulas*, G. Woan, CUP (2000). See also Reference 7.

2.2.1   sp hybridization

Mathematically, sp hybridization is the simplest case of orbital hybridization in carbon. We can begin by defining the hybridized wave functions $\Psi(sp)_{a,b}$ as linear combinations of the $\Psi(2s)$ and $\Psi(2p_x)$ atomic wave functions with coefficients C_{1-4} to be determined:

$$\Psi\left(sp\right)_a = C_1\Psi(2s) + C_2\Psi\left(2p_x\right) \qquad (2.5)$$

$$\Psi\left(sp\right)_b = C_3\Psi(2s) + C_4\Psi\left(2p_x\right)$$

The $\Psi(sp)_{a,b}$ wave functions must be orthonormal. Assuming that the $\Psi(2s)$ and $\Psi(2p_x)$ wave functions are normalized, the following values for the coefficients C_{1-4} ensure that the $\Psi(sp)_{a,b}$ wave functions are orthonormal as required:

$$C_1 = C_2 = C_3 = \frac{1}{\sqrt{2}}; \quad C_4 = -\frac{1}{\sqrt{2}} \qquad (2.6)$$

We have defined C_4 as negative here. However, we can see by inspection that whichever coefficient we define as negative, hybridized orbitals in which the electron clouds have the same spatial distribution will result – that is, $|\Psi|^2$ (the only physical observable) will be unaffected.

What do the $\Psi(sp)_{a,b}$ hybridized orbitals look like? We can visualize them by drawing schematic diagrams of the wave functions for the unhybridized orbitals $\Psi(2s)$ and $\Psi(2p_x)$ and superimposing them (Figure 2.4).

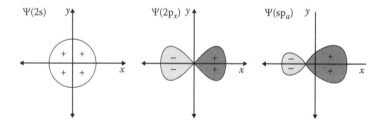

FIGURE 2.4
Schematic diagram showing the mixing, or hybridization, of carbon 2s and $2p_x$ orbitals to form the sp-hybridized orbitals (Equations 2.5 and 2.6). The addition of the symmetric $\Psi(2s)$ wave function to the antisymmetric $\Psi(2p_x)$ wave function results in hybridized wave functions $\Psi(sp)_a$ (pictured) and $\Psi(sp)_b$ in which the electron is far more likely to be found on one side of the atom, ready to form an atomic bond with an adjacent atom on that side. The spatial shape of the wave functions shown here is a simplification (albeit a commonly used one [12] of Chapter 1 and [1,8]). In actual fact the $\Psi(2s)$ wave function is slightly more complex and differs somewhat from the form it would take in a hydrogenic atom [5,9]. The simplified form we use here, however, does permit a good understanding of the underlying physics.

As shown in Figure 2.4, sp hybridization results in hybridized orbitals with an ideal shape to form molecular orbitals and strong interatomic bonds along the x-axis. We can complete our understanding of sp hybridization by considering a hypothetical one-dimensional chain of carbon atoms, carbyne. Each interatomic bond in carbyne is formed primarily from a bond called a σ-bond. The σ-bond is a molecular orbital formed from the $\Psi(sp)_a$ wave function of the atom to the left of the bond and the $\Psi(sp)_b$ wave function of the atom to the right of the bond. As illustrated in Figure 2.5, the large overlap between these wave functions ensures that the electron in this molecular orbital is very likely to be found between the two atoms and hence an extremely strong bond will be formed.

With the x-axis defined as the axis along which the carbyne chain lies, and the $\Psi(2p_x)$ orbital the only one of the $\Psi(2p_{x,y,z})$ orbitals taking part in hybridization, what happens to the other two valence electrons, in the $\Psi(2p_y)$, $\Psi(2p_z)$ orbitals? These are the electrons likely to be found along the y and z axes. Molecular orbitals (and hence interatomic bonds) also form from these (non-hybridized) atomic orbitals. However, because the $\Psi(2p_y)$, $\Psi(2p_z)$ wave functions are not oriented along the axis of the bond, the degree of overlap between the wave functions is much lower and the interatomic bonds formed are much weaker. These are called π-bonds. Figure 2.5 illustrates

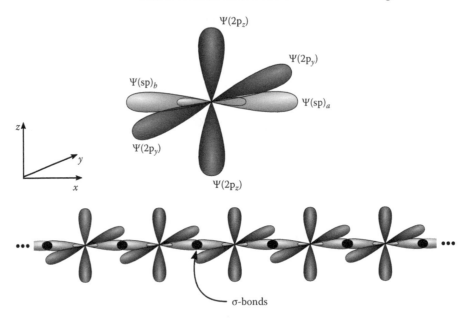

FIGURE 2.5
Schematic diagram showing formation of σ-bonds in the hypothetical one-dimensional carbyne chain of carbon atoms. The strong σ-bonds (labelled, shaded in black) form from the $\Psi(sp)_{a,b}$ hybridized orbitals with electron clouds oriented along the x-axis; hence, substantial overlap between the wave functions in adjacent atoms and formation of a strong bond. The weak π-bonds form due to the very limited overlap between the wave functions of electrons in the unhybridized $\Psi(2p_y)$, $\Psi(2p_z)$ orbitals.

the orientation of the relevant wave functions, σ-bonds and π-bonds in the carbyne chain.

The structure we consider here, an arbitrarily long one-dimensional chain of carbon atoms where adjacent atoms are linked by a σ-bond and two π-bonds, has not been observed in nature and has not been synthesized. Were it to actually exist, it would have some quite exceptional properties. The electrons in the π-bonds would move along this one-dimensional chain with comparable speed to the flow of electrons in graphene, and the σ-bonds here would be stronger than those in both graphene and diamond (this will become clear following Sections 2.2.2 and 2.2.3). It is perhaps because of this that many researchers over the past few decades have succeeded in convincing themselves that they have synthesized or discovered carbyne chains and relatively recent texts describe their existence as scientific fact.

For years, it was believed that carbyne chains of carbon atoms had been discovered in the Ries impact crater in Bavaria (where they had supposedly formed under high temperature and pressure during the impact) [10] and had also been synthesized in the laboratory by dehydrogenation of acetylene [11,12]. However, many of the samples were later re-examined and the signals attributed to carbyne were found to originate from silicate impurities [13]. Moreover, recent research has attempted to synthesize carbynes in a careful and reproducible manner by performing reactions to progressively lengthen the carbyne chain whilst keeping large functional groups attached to the ends of the chain to prevent it from coming into contact with anything it might react with. Using this technique, the longest free-standing carbyne chain synthesized to date is 44 atoms long [14]. Longer chains could not be stabilized as they were simply too reactive. This is not surprising; two out of the four bonding electrons in carbyne are in very weak π-bonds so we would expect the material to be extremely reactive. Laboratories attempting to synthesize long chains of carbyne have had some accidents on account of this, and all claims that carbyne exists and is stable as an arbitrarily long chain of carbon atoms are now thoroughly debunked [15].

A good example of sp-bonded carbon existing in nature is acetylene (C_2H_2, H–C–C–H). Even acetylene is quite reactive. In acetylene a triple bond (one σ-bond + two π-bonds) exists between the adjacent carbon atoms and a single σ-bond joins each carbon atom to a hydrogen atom. Carbon does, however, exist in nature in sp^2-bonded and sp^3-bonded forms. These are graphene/graphite and diamond, respectively.

2.2.2 sp^2 hybridization

In sp^2 hybridization, hybridized orbitals are formed from the $\Psi(2s)$ and $\Psi(2p_x)$, $\Psi(2p_y)$ wave functions, whilst the electron in the $\Psi(2p_z)$ orbital does not take part. sp^2-hybridized carbon therefore forms a layered structure where carbon atoms are joined by strong σ-bonds, but only within the xy plane: graphene or graphite. We start with the following expressions for the three normalized hybridized orbital wave functions:

$$\Psi\left(sp^2\right)_a = C_1 \Psi\left(2s\right) - \sqrt{1-C_1^2}\,\Psi\left(2p_y\right) \tag{2.7}$$

$$\Psi\left(sp^2\right)_b = C_2 \Psi\left(2s\right) + \sqrt{1-C_2^2}\left[\frac{\sqrt{3}}{2}\Psi\left(2p_x\right) + \frac{1}{2}\Psi\left(2p_y\right)\right]$$

$$\Psi\left(sp^2\right)_c = C_3 \Psi\left(2s\right) + \sqrt{1-C_3^2}\left[-\frac{\sqrt{3}}{2}\Psi\left(2p_x\right) + \frac{1}{2}\Psi\left(2p_y\right)\right]$$

Just as with sp hybridization, we expect the mixing of the symmetric $\Psi(2s)$ orbital and the antisymmetric $\Psi(2p_{x,y})$ orbitals to result in hybridized orbitals which are lopsided, with the electron likely to be found in a specific region on one side of the atom. With both the $\Psi(2p_x)$ and $\Psi(2p_y)$ orbitals available to take part in hybridization, the hybridized orbitals will be wave functions in which the electron is likely to be found at three different locations around the nucleus in the xy plane. We expect the lowest energy configuration to be that in which the electrons are as far away from each other as possible: angles of 120° between the different electron clouds (and hence, eventually, between the interatomic bonds).

For simplicity we start with one hybridized orbital oriented along the y-axis, $\Psi(sp^2)_a$ in Equation 2.7. We then specify, for $\Psi(sp^2)_b$ and $\Psi(sp^2)_c$, the proportions of the unhybridized $\Psi(2p_x)$ and $\Psi(2p_y)$ orbitals which contribute to each hybridized orbital. This is to ensure the electron clouds forming the hybridized orbitals sit as far away from each other as possible (at 120° separation). This leaves us with three coefficients to determine: C_1–C_3. These can be determined by referring to the Pauli exclusion principle: Each of the unhybridized orbitals must still be occupied by exactly one electron. Thus,

$$C_1 = C_2 = C_3 = \frac{1}{\sqrt{3}} \tag{2.8}$$

Similar to the case of sp hybridization (Figure 2.4), we can draw a schematic diagram to illustrate the lopsided nature of the hybridized orbitals (Figure 2.6). Due to each hybridized orbital not receiving such a high proportion of the symmetric $\Psi(2s)$ orbital, the hybridized wave function is not quite so lopsided as in the case of sp hybridization. Therefore, the amount of overlap between the hybridized atomic orbitals

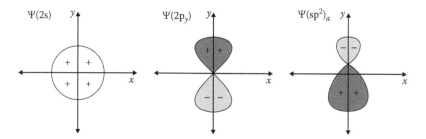

FIGURE 2.6
Schematic diagram showing the mixing, or hybridization of carbon orbitals to form the $\Psi(sp^2)_a$ hybridized orbital (Equations 2.7 and 2.8). The addition of the symmetric $\Psi(2s)$ orbital to the antisymmetric $\Psi(2p_y)$ orbital results in a hybridized wave function $\Psi(sp^2)_a$ in which the electron is far more likely to be found on one side of the atom, ready to form an atomic bond with an adjacent atom on that side.

that form the molecular orbital is not so great as in the case of sp hybridization. This results in a slightly weaker, but still very strong, bond. In Section 2.3, we elaborate on this by comparing the bond energies of the sp, sp^2 and sp^3 bonds and discuss in detail the results of making different linear combinations of generic symmetric and antisymmetric wave functions.

We will now take a look at the graphene lattice formed from pure sp^2-hybridized carbon (Figure 2.7). Whilst the hypothetical carbyne chain of sp-hybridized carbon is a chain of identical atoms, the primitive unit cell of the graphene lattice consists of two non-identical atoms (A and B): The B atom has the same hybridized orbitals as the A atom, but rotated through 180°. We can therefore write down expressions (Equation 2.9) for the hybridized wave functions for the B atom (Equations 2.7 and 2.8 gave the expressions for the A atom):

$$\Psi\left(sp^2\right)_a = \frac{1}{\sqrt{3}}\Psi\left(2s\right)+\sqrt{\frac{2}{3}}\Psi\left(2p_y\right) \qquad (2.9)$$

$$\Psi\left(sp^2\right)_b = \frac{1}{\sqrt{3}}\Psi\left(2s\right)+\sqrt{\frac{2}{3}}\left[\frac{\sqrt{3}}{2}\Psi\left(2p_x\right)-\frac{1}{2}\Psi\left(2p_y\right)\right]$$

$$\Psi\left(sp^2\right)_c = \frac{1}{\sqrt{3}}\Psi\left(2s\right)-\sqrt{\frac{2}{3}}\left[\frac{\sqrt{3}}{2}\Psi\left(2p_x\right)+\frac{1}{2}\Psi\left(2p_y\right)\right]$$

The final (but very important) matter we must discuss relating to sp^2 hybridization is the fate of the electron in the $\Psi(2p_z)$ orbital, the orbital which does not take part in hybridization. Just like the $\Psi(2p_y)$ and $\Psi(2p_z)$ orbitals in

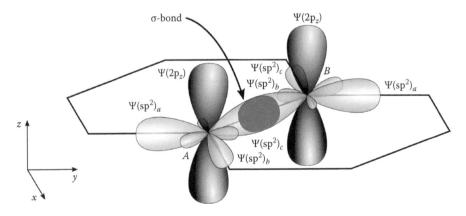

FIGURE 2.7
Interatomic bonding in the graphene lattice (sp^2 bonding). The σ-bonds are formed by the overlap of hybridized orbitals along the bond axis, and are responsible for the strength of the sp^2 C–C bond. The π-bonds are formed by the overlap of the non-hybridized $\Psi(2p_z)$ atomic orbitals. They contribute little to the total strength of the C–C bond but are responsible for graphene's remarkable electronic properties.

sp hybridization, the electrons in the $\Psi(2p_z)$ orbitals in sp^2 hybridization form weak π-bonds with neighbouring carbon atoms (Figure 2.7) and, in the case of graphite, also the weak van der Waals bonds which hold the graphene layers together. As we shall see in Chapter 3, whilst the electrons in the hybridized orbitals form the bonds which are primarily responsible for holding the graphene sheet together, it is the electrons in the π-bonds which are responsible for the extremely high carrier mobility and conductivity of graphene.

2.2.3 sp^3 hybridization

The third form of hybridization exhibited by the carbon atom is sp^3 hybridization. In sp^3 hybridization, all of the valence electron orbitals: $\Psi(2s)$, $\Psi(2p_x)$, $\Psi(2p_y)$, $\Psi(2p_z)$, contribute to the formation of the hybridized orbitals. There are therefore four hybridized orbitals, forming molecular orbitals for σ-bonds to adjacent atoms with a three-dimensional tetrahedral structure: diamond. Since all atomic orbitals take part in hybridization there are no π-bonds, only σ-bonds formed from hybridized orbitals.

The $\Psi(2s)$ and $\Psi(2p_{x,y,z})$ atomic orbitals combine to give the (normalized) sp^3-hybridized orbitals as shown in Equation 2.10. In the case of sp^3-hybridization, we expect an isotropic material (as diamond is observed to be), and to achieve this the bonds pointing in different directions are identical. We also expect a tetrahedral structure with bond angles of 109.5° to keep adjacent electron clouds as far apart as possible.

The coefficients in Equation 2.10, the expressions for the sp^3-hybridized orbitals, are chosen to ensure this:

$$\Psi\left(sp^3\right)_a = \frac{1}{2}\left[\Psi(2s) + \Psi(2p_x) + \Psi(2p_y) + \Psi(2p_z)\right] \quad (2.10)$$

$$\Psi\left(sp^3\right)_b = \frac{1}{2}\left[\Psi(2s) - \Psi(2p_x) - \Psi(2p_y) + \Psi(2p_z)\right]$$

$$\Psi\left(sp^3\right)_c = \frac{1}{2}\left[\Psi(2s) - \Psi(2p_x) + \Psi(2p_y) - \Psi(2p_z)\right]$$

$$\Psi\left(sp^3\right)_d = \frac{1}{2}\left[\Psi(2s) + \Psi(2p_x) - \Psi(2p_y) - \Psi(2p_z)\right]$$

Figure 2.8 illustrates the spatial orientation of the sp^3 bonds, and the diamond structure that results. From the crystallographic point of view, the structure can be described as face-centred cubic with two atoms in each unit cell and the lattice vectors joining a corner of the cube to the centres of the three adjacent faces. Each sp^3-bond consists of a single σ-bond but (unlike sp and sp^2 hybridization) no π-bond, because all atomic orbitals have taken part in hybridization.

In addition to diamond, there are many other systems in nature involving sp^3-bonding. In methane (CH_4) the central carbon atom is bonded to each hydrogen atom via an sp^3 σ-bond, and crystalline silicon at ambient conditions

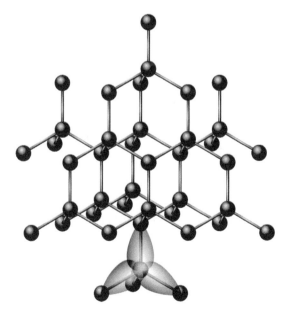

FIGURE 2.8
sp^3-hybridized orbitals in carbon, incorporated into the diamond lattice.

also exhibits sp³-bonding (where the hybridized orbitals are formed from the Ψ(3s) and Ψ(3p) atomic orbitals).

Most readers will have been motivated to read this book by the myriad of potential applications for sp²-bonded carbon (graphene), but it is worthwhile to reflect on the diverse panoply of applications which have existed for sp³-bonded carbon for many decades. Diamond cutting tools play a key role in mechanical engineering for applications such as oil drilling; without the diamond anvil high pressure cell, we would not be able to study the earth's interior in the laboratory, and other applications such as heatsinks and high-performance bearings abound. Given its many industrial and cosmetic applications, the diamond structure has been very well characterized over the decades.

2.3 Lengths and bond energies of sp, sp² and sp³ bonds

The sp, sp² and sp³ bonds are all molecular orbitals formed from hybridized atomic orbitals. The overlap between the hybridized orbitals of electrons in adjacent atoms is large because the hybridized atomic orbital consists of a lopsided electron wave function – the electron is more likely to be found on the side of the atom where the bond is being formed (Figures 2.4 and 2.6).

The unhybridized atomic wave functions Ψ(2s), Ψ(2p$_{x,y,z}$) on the other hand, are all symmetric or antisymmetric about the nucleus, resulting in an electron which is equally likely to be found on either side of the nucleus. The reason why the hybridized orbital results in the electron being more likely to be found on one side of the nucleus is because it is formed by combining a symmetric wave function (the Ψ(2s)) with one or more antisymmetric Ψ(2p$_{x,y,z}$) wave functions. The lengths and bond energies of sp, sp² and sp³ bonds can be understood in terms of what proportion of the hybridized orbital comes from the symmetric Ψ(2s) wave function and what proportion comes from antisymmetric Ψ(2p$_{x,y,z}$) wave functions. It is therefore worthwhile to study and understand the mixing of symmetric and antisymmetric wave functions.

2.3.1 Addition of symmetric and antisymmetric wave functions

Referring to Figure 2.9, we can see the electron probability density distribution which arises from addition of example symmetric and antisymmetric wave functions. When we form a hybrid wave function from different proportions of the symmetric and antisymmetric wave functions (which are otherwise identical), the result is an electron density distribution which is lopsided, that is, the electron is more likely to be found on one side of the nucleus (positive x in Figure 2.9). The most lopsided probability distribution (case (c)) is that where we mix equal proportions of the symmetric and antisymmetric wave functions to create the hybrid wave function. Mixing the symmetric and antisymmetric wave functions in

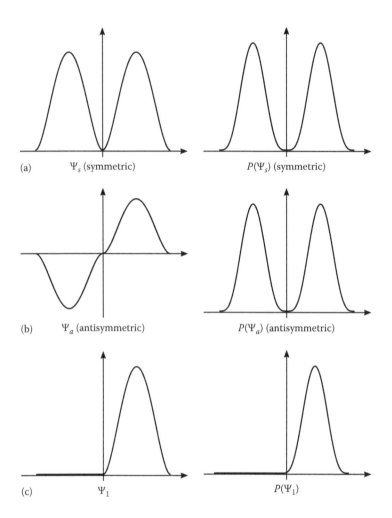

(a) Ψ_s (symmetric)

$P(\Psi_s)$ (symmetric)

(b) Ψ_a (antisymmetric)

$P(\Psi_a)$ (antisymmetric)

(c) Ψ_1

$P(\Psi_1)$

FIGURE 2.9

Examples of the wave functions and electron probability distributions resulting from the addition, or mixing, of different wave functions. (a) A wave function (Ψ_s) which is symmetric about the origin results in a probability density which is also symmetric about the origin. (b) A wave function (Ψ_a) which is antisymmetric about the origin results in a probability density which is also symmetric about the origin since $P(x) = |\Psi(x)|^2$. (c) A wave function $\left(\Psi_1 = \dfrac{1}{\sqrt{2}}[\Psi_s + \Psi_a] \right)$ consisting of equal proportions of the symmetric (Ψ_s) and antisymmetric (Ψ_a) wave functions results in the particle which can only be found in the positive x region.   (*Continued*)

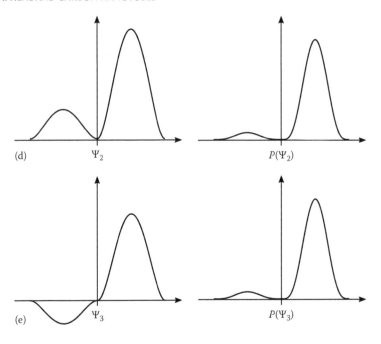

(d) Ψ_2 $P(\Psi_2)$

(e) Ψ_3 $P(\Psi_3)$

FIGURE 2.9 (*Continued*)
Examples of the wave functions and electron probability distributions resulting from the addition, or mixing, of different wave functions. (d) and (e) Mixing a large proportion of the symmetric wave function with a small proportion of the antisymmetric wave function (or vice versa) $\left(\Psi_2 = \dfrac{\sqrt{3}}{2}\Psi_s + \dfrac{1}{2}\Psi_a \text{ or } \Psi_3 = \dfrac{1}{2}\Psi_s + \dfrac{\sqrt{3}}{2}\Psi_a\right)$ results in the particle being more likely to be found in the positive x region but (unlike (c)) still has some probability of being found in the region of negative x. All wave functions are normalized.

unequal proportions always results in a probability distribution ((d) or (e)) which is less lopsided than case (c).

In the example shown in Figure 2.9, we start with two wave functions which are identical except for their symmetric and antisymmetric nature. In the carbon atom, it is not quite so clear-cut* because we mix a symmetric wave function ($\Psi(2s)$) with antisymmetric wave functions $\left(\Psi\left(2p_{x,y,z}\right)\right)$ which are not otherwise identical to the $\Psi(2s)$. But nonetheless, the basic conclusion is the same: We get the most lopsided electron probability density when we mix equal proportions of the symmetric and antisymmetric wave functions. This is the case of sp bonding (Equations 2.5 and 2.6). As a result, the sp bond is the shortest out of the three C–C bonds – the electron is most likely to be found between the two carbon nuclei, and the nuclei are thus pulled closer together than is the case with sp^2 and sp^3 bonding. The sp bond also requires

---

* Compare case (c) in Figure 2.9 with Figure 2.4.

TABLE 2.1

Accepted bond lengths and energies for the sp, sp^2 and sp^3 C–C bonds

Bond	Length (pm)	Energy (kJ mol^{-1})
sp C–C	121	837
sp^2 C–C	134	611
sp^3 C–C	154	347

Source: Kotz, J.C. and Purcell, K.F., *Chemistry & Chemical Reactivity*, Saunders College Publishing, 1987.

Note: These are example values, the actual lengths and energies do vary slightly between different materials.

the most energy to break, because the fact that the electron is most likely to be found between the two nuclei being bonded results in a molecular bonding orbital formed by a larger overlap between atomic orbitals.

In both sp^2 and sp^3 bonding, the antisymmetric $\Psi(2p_{x,y,z})$ wave functions make up a greater proportion of the hybridized orbital than the symmetric $\Psi(2s)$ wave function (Equations 2.9 and 2.10). The hybridized orbital can therefore be considered to have more p character than s character, analogous to case (e) in Figure 2.9. The sp^2 orbital has more p character than the sp orbital, and the sp^2 bond is therefore longer and weaker than the sp bond. The sp^3 orbital has even more p character, resulting in a bond which is longer and weaker still: but still an extremely strong bond.

Table 2.1 shows accepted generic bond lengths and energies for the sp, sp^2 and sp^3 C–C bond. Such values have existed in the literature for many decades prior to the discovery of graphene; they are obtained from measurements and experiments on hydrocarbons and other organic materials.

The trend in bond length and energy as we move from sp through sp^2 to sp^3 bonding is exactly what we expect as a result of our discussion earlier regarding the formation of hybridized atomic orbitals from the addition of symmetric and antisymmetric wave functions.

2.4 What if the bonding geometry does not allow pure sp, sp^2 or sp^3 hybridization?

Thus far, we have discussed sp, sp^2 and sp^3 hybridization in a manner appropriate to materials such as graphene and benzene (sp^2), or diamond and methane (sp^3). In these materials the system is free to take the optimum bond angles that give the largest possible overlap between the adjacent hybridized atomic orbitals formed through pure sp^2 and sp^3 hybridization. The system also exhibits the largest possible separation between the electron clouds forming adjacent bonds. In the graphene and benzene systems, the bonds have the freedom to form at 120° bond angles in the xy plane, and in the diamond and methane systems the bonds have the freedom to form at 109.5° bond angles in the tetrahedral structure. The values given in Table 2.1 for the bond lengths and energies apply only to systems such as this.

But what if the bonds cannot form at the optimum angles for sp, sp^2 or sp^3 hybridization? For instance, what if we take a graphene sheet and roll it up to form a single-walled carbon nanotube? In that case, the bonds are no longer able to form at 120° in the same plane so pure sp^2 hybridization is no longer possible. Instead, the hybridization can be described as a mixture of sp^2 and sp^3 hybridization, or $sp^{2+\eta}$ hybridization where $0 < \eta < 1$. Whilst the hybridization in single-walled nanotubes has a very small sp^3 component [17], fullerenes have a much larger sp^3 component as may be expected from the larger curvature ([12] of Chapter 1).

References

1. MJS Dewar, *The Molecular Orbital Theory of Organic Chemistry*, McGraw-Hill, New York (1969).

2. M Born and R Oppenheimer, *Annalen der Physik* **389**, 457 (1927).

3. RS Mulliken, Nobel Prize Lecture (1966).

4. PW Atkins, *Physical Chemistry*, OUP, Oxford, U.K. (1982).

5. BH Bransden and CJ Joachain, *Physics of Atoms and Molecules*, Prentice Hall, Upper Saddle River (1983).

6. AB Callear and REM Hedges, *Nature* **215**, 1267 (1967).

7. AIM Rae, *Quantum Mechanics*, IoP Publishing, Bristol (1992).

8. FA Carey and RM Giuliano, *Organic Chemistry*, McGraw-Hill, New York (2008).

9. TW Graham Solomons, *Fundamentals of Organic Chemistry*, Wiley-VCH, Weinheim (1982).

10. A El Goresy and G Donnay, *Science* **161**, 363 (1968).

11. VI Kasatochkin et al., *Carbon* **11**, 70 (1973).

12. VI Kasatochkin et al., *Dokl. Acad. Nauk USSR* **201**, 1104 (1971).

13. PPK Smith and PR Buseck, *Science* **216**, 984 (1982).

14. WA Chalifoux and RR Tykwinski, *Nat. Chem.* **2**, 967 (2010).

15. H Kroto, Carbyne and other myths about carbon. *Chem. World* **7**, 1014913 (November 2010).

16. JC Kotz and KF Purcell, *Chemistry & Chemical Reactivity*, Saunders College Publishing (1987).

17. A Kleiner and S Eggert, *Phys. Rev. B* **64**, 113402 (2001).

3 Electronic Dispersion Relation of Graphene

3.1 Some introductory remarks on the application of tight-binding theory to graphene

In general terms, there are two approaches to predicting electronic dispersion relations (band structure) of solids: nearly free electron theory and tight-binding theory. In nearly free electron theory, we make the simplifying assumption that the valence electrons are nearly free particles. The effect of the periodic potential due to the ionic cores of the atoms on the electron wave function is small. In tight-binding theory, on the other hand, we assume that the wave function of the electron in the system can be approximated by a linear combination of the wave functions of electrons bound to individual atoms in the system (atomic orbitals), wave functions which overlap by some small amount.

We will begin with a few general remarks about tight-binding theory.

The first remark regards the assumption that the wave function of the electron in the system can be described by a linear combination of the wave functions of electrons bound to individual atoms in the system with a small overlap. This is, of course, the same approximation as we made in the last chapter when we applied molecular orbital theory to graphene. So tight-binding theory is simply the application of molecular orbital theory to calculate the electronic dispersion relation.

Second, when we assume that the electron wave function can be given by a linear combination of atomic orbitals which overlap by a small amount, we are not assuming that the electrons cannot move easily from one atom to the next. As we will see in this chapter, graphene is a good example of a system for which tight-binding theory works well, but in which (some of) the electrons can move through the crystal with astonishing speed. Arguably the name "tight-binding theory" is misleading in this context.

In this chapter, we apply tight-binding theory to graphene. Following on from Chapter 2, we will see that we can consider the bonding orbitals as the valence bands in graphene's electronic dispersion relation and that we can consider the antibonding orbitals as the conduction bands. For the case of the π-electrons, we will see that the valence and conduction bands meet; it takes very little energy to excite these electrons from the valence band (bonding orbital) to the conduction band (anti-bonding orbital). Hence, graphene is a good conductor as a result of the weakness of the π-bonds.

In the case of the σ-electrons, however, there is a large energy gap between the valence band (bonding orbitals) and the conduction band (anti-bonding orbitals). These electrons cannot easily travel through the crystal as they are bound in strong bonds.

3.2   Application of molecular orbital theory to the π-electrons in graphene

In Chapter 2, we applied molecular orbital theory to understand the formation of the σ-bonds in graphene. We can write down an expression for the molecular orbital in which these electrons sit quite easily. For instance (referring to Figure 2.7), the molecular orbital forming the σ-bond between atoms A and B would be written as

$$\Psi(\sigma) = \frac{1}{\sqrt{2}}\left[\Psi_A\left(sp^2\right)_b \pm \Psi_B\left(sp^2\right)_b\right] \tag{3.1}$$

where $\Psi_{A,B}(sp^2)_b$ are the relevant normalized hybridized orbitals for the A and B atoms, respectively—given by Equations 2.7 through 2.9.

In this chapter, however, we are primarily interested in the molecular orbitals and electronic dispersion relation due to the electrons which do not take part in hybridization: Those that sit in the molecular orbitals responsible for the π-bonds, formed from overlapping $2p_z$ atomic orbitals. The $\Psi(2p_z)$ wave function points up and down along the z-axis, perpendicular to the direction of the bonds. It therefore overlaps equally with the equivalent wave functions in all three neighbouring atoms and forms a molecular orbital due to the overlap with all three of these wave functions.

We can therefore apply molecular orbital theory to the formation of the π-bonds by considering a graphene primitive unit cell defined differently to that in Chapter 2: This time formed from an A atom at the origin and one-third of each of the B atoms (Figure 3.1). The two atoms that form this primitive unit cell contribute in total two $2p_z$ electrons which will both sit in the molecular orbital with normalized wave function $\Psi(\pi)$ defined in the following equation:

$$\Psi(\pi) = \frac{1}{\sqrt{2}}\Psi_A\left(2p_z\right) \pm \frac{1}{\sqrt{6}}\left[\Psi_{B1}\left(2p_z\right) + \Psi_{B2}\left(2p_z\right) + \Psi_{B3}\left(2p_z\right)\right]$$

$$\tag{3.2}$$

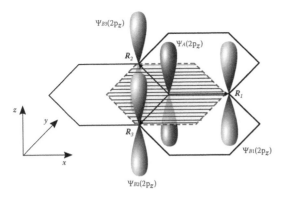

FIGURE 3.1

Graphene unit cell (marked with horizontal shading) consisting of a single A atom at the origin, and one-third of each of the three adjacent B atoms at locations R_1–R_3. The tight-binding wave function (Equation 3.2) is a linear combination of the $\Psi_A(2p_z)$ wave function (dark shading) and the $\Psi_{B1-3}(2p_z)$ wave function (light shading).

In Equation 3.2, we have formed a molecular orbital by considering the overlap between the $\Psi(2p_z)$ wave function in the A atom and its three nearest neighbour B atoms. It is therefore useful to write the expression as follows:

$$\Psi(\pi) = \frac{1}{\sqrt{2}} \Psi_A(2p_z) \pm \frac{1}{\sqrt{6}} \sum_n \Psi_{Bn}(2p_z) \qquad (3.3)$$

where the summation is taken over all three of the A atom's nearest neighbour B atoms.

3.3 Simple tight-binding calculation for π-electrons in graphene

We can apply tight-binding theory in order to calculate the energy $E(k)$ of the π electron in the molecular orbital specified in Equation 3.3, but propagating through the graphene lattice with wavevector k. The wavevector enters the calculation because it determines the phase shift between the values of the electron wave function at adjacent atomic sites. To find the energy we apply the Hamiltonian operator $\hat{H}$ to the function $\Psi(\pi)$. We then pre-multiply the expression by $\Psi_A^*(2p_z)$, and integrate over all space. We obtain the expression:

$$\hat{H}\Psi(\pi) = E(k)\Psi(\pi) \qquad (3.4)$$

$$\frac{1}{\sqrt{2}}\int\Psi_A^*(2p_z)\hat{H}\Psi_A(2p_z)d\tau \pm \frac{1}{\sqrt{6}}\sum_n\int\Psi_A^*(2p_z)\hat{H}\Psi_{Bn}(2p_z)d\tau$$

$$=\frac{E(k)}{\sqrt{2}}\int\Psi_A^*(2p_z)\Psi_A(2p_z)d\tau \pm \frac{E(k)}{\sqrt{6}}\sum_n\int\Psi_A^*(2p_z)\Psi_{Bn}(2p_z)d\tau$$

We can separately evaluate each term in this expression. The first integral simply gives the energy ε_{2p} of an electron in the $\Psi(2p_z)$ orbital of a carbon atom in the graphene lattice, shown in the following equation:

$$\int\Psi_A^*(2p_z)\hat{H}\Psi_A(2p_z)d\tau = \epsilon_{2p} \tag{3.5}$$

The third integral is equal to one, as we have started with normalized wave functions. The second and fourth terms, however, warrant more detailed consideration. Both are overlap integrals between the $\Psi(2p_z)$ wave functions for the A and B atoms. For symmetry reasons, we expect the magnitude of this overlap to be the same for all three nearest neighbour B atoms, but since we are interested in electrons propagating through the lattice with a non-zero wavevector k, we must account for the fact that the wave function's value will undergo a phase shift between the sites of the A and B atoms.* We can therefore write

$$\frac{1}{\sqrt{3}}\sum_n\int\Psi_A^*(2p_z)\hat{H}\Psi_{Bn}(2p_z)d\tau = t\sum_n e^{ik\cdot R_n} \tag{3.6}$$

$$\frac{1}{\sqrt{3}}\sum_n\int\Psi_A^*(2p_z)\Psi_{Bn}(2p_z)d\tau = s\sum_n e^{ik\cdot R_n}$$

for all three nearest-neighbour B atoms (n = 1–3). s and t are both overlap integrals which have the same value for n = 1–3.

With the A atom at the origin, the B atoms are located at the following locations (Figure 3.1):

$$R_1 = \begin{pmatrix} \dfrac{a}{\sqrt{3}} \\ 0 \end{pmatrix} \tag{3.7}$$

$$R_2 = \begin{pmatrix} -\dfrac{a}{2\sqrt{3}} \\ \dfrac{a}{2} \end{pmatrix}$$

---

* Our allowance here for the fact that the wave function's value can undergo a phase shift between adjacent atomic sites is an allowance for the only change in the wave function permitted by Bloch's theorem (Section B.1).

$$R_3 = \begin{pmatrix} -\dfrac{a}{2\sqrt{3}} \\ -\dfrac{a}{2} \end{pmatrix}$$

where a is the scalar value of the graphene lattice vector (see Equation 1.1). We can therefore substitute into Equation 3.4 to write

$$\varepsilon_{2p} \pm tf(\mathbf{k}) = E(\mathbf{k}) \pm E(\mathbf{k})sf(\mathbf{k}) \tag{3.8}$$

where the factor $f(\mathbf{k})$ accounts for the phase change in the electron wave function as we move from the A atom site to the B atom sites:

$$f(\mathbf{k}) = \sum_n e^{i\mathbf{k}\cdot R_n} = e^{\frac{ik_xa}{\sqrt{3}}} + 2e^{-\frac{ik_xa}{2\sqrt{3}}}\cos\left(\frac{k_ya}{2}\right) \tag{3.9}$$

We can obtain, from Equation 3.9,*

$$E(\mathbf{k}) = \frac{\varepsilon_{2p} \pm tw(\mathbf{k})}{1 \pm sw(\mathbf{k})} \tag{3.10}$$

where $w(\mathbf{k}) = |f(\mathbf{k})|$

$$w(\mathbf{k}) = |f(\mathbf{k})| = \sqrt{1 + 4\cos\frac{\sqrt{3}k_xa}{2}\cos\frac{k_ya}{2} + 4\left(\cos\frac{k_ya}{2}\right)^2} \tag{3.11}$$

A detailed calculation taking into account the nature of the $\Psi(2p_z)$ wave functions can provide values for ε_{2p}, s and t. Various authors have considered this from the theoretical point of view over the decades both prior to and since the discovery of graphene in 2004 ([6–8,12] of Chapter 1 and [1–3]). We shall use the values from References 8 and 12 of Chapter 1: $\varepsilon_{2p} = 0$, $t = -3.033$ eV, $s = 0.129$. These values were obtained theoretically prior to the discovery of graphene in 2004 but have since been found to accurately predict the velocity of the mobile electrons in graphene (as determined by studying the cyclotron motion of electrons in graphene ([31,32] of Chapter 1) and performing angle-resolved photoelectron spectroscopy (ARPES) on graphene [4–6]). t always has a negative value, and by setting $\varepsilon_{2p} = 0$, we are simply setting the arbitrary zero energy point at the Fermi level. We will set $\varepsilon_{2p} = 0$ for the remainder of this book.

We can plot (Figure 3.2) the π-electron dispersion relation for graphene using Equation 3.10. This is the figure that you

---

* Via the intermediate steps $\left[t - sE(\mathbf{k})\right]^2 f(\mathbf{k})f^*(\mathbf{k}) = \left[E(\mathbf{k}) - \varepsilon_{2p}\right]^2$, then $\pm\left[t - sE(\mathbf{k})\right]w(\mathbf{k}) = E(\mathbf{k}) - \varepsilon_{2p}$.

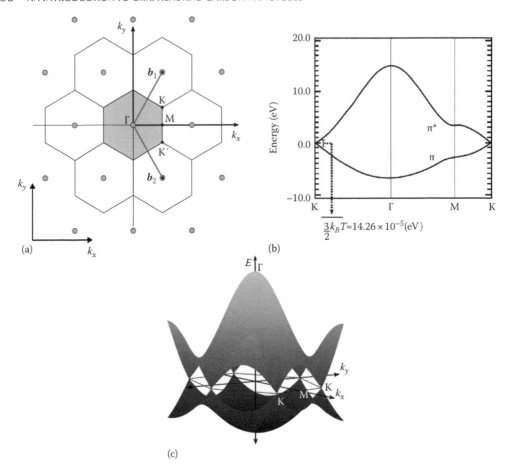

FIGURE 3.2
(a) Graphene reciprocal lattice with first Brillouin zone shaded, reciprocal lattice vectors and high symmetry points marked. (b) Plot of the energy dispersion relation for the π-electrons in graphene along high symmetry directions. Thermal energy available at 300 K is labelled. (c) Full energy dispersion relation. The conduction band (upper) and the valence band (lower) meet at a point known as the K point; making graphene a zero-bandgap semiconductor.

first came across in Chapter 1. Whilst the simple tight-binding method presented in this chapter has given us the correct answer, its ability to tell us about the underlying physics is limited; in particular, Bloch's theorem has been mentioned only in passing, and, moreover, we have not derived the tight-binding Hamiltonian matrix for the electrons in graphene. This is crucial to understanding the parallels between the behaviour of the electrons in graphene and massless relativistic fermions. The more advanced tight-binding methodology presented in Appendix B is more complex mathematically, but it overcomes these shortcomings, and any reader with a serious interest in the electronic properties of graphene is strongly recommended to familiarize themselves with it.

It is often useful to use a simple approximation to the energy dispersion relation contained in Equations 3.10 and 3.11. We can obtain this by performing a binomial expansion on Equation 3.10 for the case of $\varepsilon_{2p} = 0$:

$$E(k) = \pm \frac{tw(k)}{1 \pm sw(k)} \approx \pm tw(k)\left(1 \mp sw(k) + \left(sw(k)\right)^2 + \cdots\right)$$

In the case of small s (small amount of overlap between electron wave functions from adjacent atoms) and also the case of small $w(k)$ (this is the case for the dispersion relation close to the K point where $w(k) = 0$), it is a good approximation to keep only the first-order term in the binomial expansion and write the commonly used approximation for the energy dispersion relation:

$$E(k) \approx \pm tw(k) \tag{3.12}$$

3.4   Valence and conduction bands in graphene

The $\pm$ signs in Equation 3.10 allow a positive and a negative value for $E(k)$. These correspond to the bonding orbital and valence band (negative $E(k)$), and to the anti-bonding orbital and conduction band (positive $E(k)$). See Figure 3.2. The bonding molecular orbital for the primitive unit cell defined in Section 3.2 can be occupied by two electrons with opposing spins. It is hence full as each carbon atom contributes one electron to the orbital. Therefore, the valence band is also full. The anti-bonding orbital is the conduction band, which is empty. However, the valence band (bonding orbitals) and conduction band (anti-bonding orbitals) meet at the K point (Figure 3.2), so at non-zero temperature, enough thermal energy is always available to excite electrons to the conduction band. Hence, graphene (just like graphite) is a good conductor of electricity in the plane of the graphene layer.

We could, if we wished, perform an equivalent tight-binding calculation for the electrons in the σ-bonds formed from the hybridized orbitals (see, for instance, Reference 12 of Chapter 1). In this case, we find that there is a substantial energy gap (around 10 eV) between the conduction and valence bands due to the fact that these electrons are held in the strong σ-bonds.

However, if we convert the thermal energy available at ambient temperature ($3/2k_BT$ at $T = 300K$) into units of electron volts (eV), we can see (Figure 3.2) that at ambient temperature we can only excite π-electrons from the valence band to the conduction band in the region very close to the K point. We shall therefore consider this region in more detail in the following section.

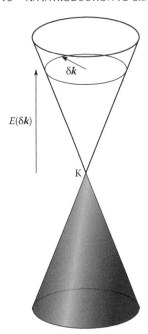

$E(\delta k)$

δk

K

FIGURE 3.3

The area close to the K point in graphene's electronic dispersion relation, with energy and wavevector measured relative to the K point. The valence band is shaded.

3.5 Massless Dirac fermions in graphene

Many readers will have read about the unique and incredible electronic properties of graphene; in particular the existence of massless electrons moving at 10^6 ms^{-1}. We are now ready to understand from first principles, using our tight-binding calculation, why electrons in graphene have these properties. Let us zoom in on the area close to the K point in the dispersion relation (Figure 3.3).

The dispersion relation appears conical in this region, and we can prove that this appearance is correct. To study the region of the dispersion relation close to the K point, we define the wavevector k as

$$k = k_K + \delta k \tag{3.13}$$

where k_K is a vector leading from the Γ point to the K point.

We can perform a binomial expansion on Equation 3.11 and make approximations to the trigonometrical functions which are valid close to the K point (i.e. small δk). We find that $w(k)$ depends only on the scalar value δk. The following expression for $w(k)$ valid in this region is obtained:

$$w(k) \approx \frac{\sqrt{3}}{2} a\delta k$$

Utilizing Equation 3.12, we can obtain

$$E(k) \approx \pm t \frac{\sqrt{3}}{2} a\delta k \tag{3.14}$$

Readers are invited to perform this derivation in the problems. The fact that $w(k)$ and $E(k)$ depend only on the scalar value of δk is the proof that the dispersion relation is conical in this region. The group velocity $v(k)$ of electrons in the crystal can be obtained from the dispersion relation ([3] of Chapter 1). In this simple case we need only consider the scalar value of k and can use the expression

$$v(k) = \frac{1}{\hbar} \left| \frac{dE(k)}{dk} \right| = \frac{\sqrt{3}a|t|}{2\hbar} \tag{3.15}$$

Using Equations 3.14, 3.15, $t = -3.033$ eV ([8,12] of Chapter 1), and $a = 2.46$ Å ([1] of Chapter 1), we can evaluate the group velocity $v(k)$ of electrons at the Fermi level* (i.e. at the K point). Since the dispersion relation is linear, the group velocity is a constant: $v(k) = v_F = 1.0 \times 10^6$ ms^{-1}.

---

* The Fermi level is the highest occupied electronic energy level at $T = 0$ K, that is, when the electrons are occupying the lowest energy levels permitted by the Pauli exclusion principle. This corresponds to the valence band being full and the conduction band being empty. The Fermi velocity is the group velocity of an electron in a quantum state at the Fermi level.

We can also understand why it is appropriate to refer to the electrons in graphene as massless. In the region close to the K point, the relationship between their energy and wavevector is linear. This is normally the case for a massless particle, for example, for a photon we write $E = \hbar kc$. For a particle with non-zero effective mass on the other hand, we would usually write $E = (\hbar^2 k^2 / 2m^*)$.

Whilst we obtained our group velocity using the results of a theoretical calculation performed years prior to the isolation of graphene, we can now compare it to experimental data in which the electron dispersion relation has been directly measured. In an ARPES experiment on graphene, we shine an extremely intense beam of soft x-rays on the graphene sample. We measure the energy and emission angle of the electrons ejected from the graphene upon absorption of the x-ray photons. From this, we can determine the energy and momentum of the ejected electrons, and of course we know the energy and momentum of the incident x-rays. We can hence, using the conservation of energy and momentum, determine the energy–momentum relation (i.e. electronic dispersion relation, as shown in Figures 3.2 and 3.3) of the electrons in the graphene layer.

In Figure 3.4, we reproduce typical ARPES data [5] on a graphene monolayer covering the region close to the K point and compare it to our theoretical prediction (Equation 3.14). We can see that the theoretical prediction (performed years prior to the discovery of graphene) is remarkably accurate. The electrons really do move at a relativistic velocity close to $v_F \approx 1.0 \times 10^6$ ms^{-1}.

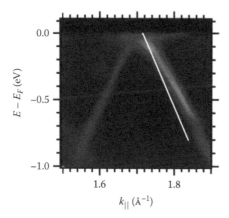

FIGURE 3.4
Comparison of theoretical prediction (white line, given by Equation 3.14) and angle-resolved photoelectron spectroscopy experimental data on the electronic dispersion relation close to the K point in graphene. (Reprinted with permission from Pletikosić, I., Kralj, M., Pervan, P., Brako, R., Coraux, J., N'Diaye, A.T., Busse, C., and Michely, T., *Phys. Rev. Lett.*, 102, 056808. Copyright 2009 by the American Physical Society.)

Why do we refer to the sections of the dispersion relation close to the K points (as shown in Figures 3.2 and 3.3) as the "Dirac cones"? It is because, if we construct a tight-binding Hamiltonian matrix for the electrons near to the K point using the linear dispersion approximation (Equation 3.14), it is identical in mathematical form to the Hamiltonian in the Dirac equation ([31,32] of Chapter 1 and [1,7,8]), the relativistic equivalent of the Schrödinger equation for the case of spin-(1/2) particles. We will see in the next chapter that the use of this method is essential to correctly predict the major anomalous features of the quantum Hall effect in graphene, and the interested reader is referred to Section B.3 for the formal derivation of the Dirac-like Hamiltonian matrix.

It is important to note (and is evident from the derivation in Section B.3) that the use of the relativistic Dirac equation to describe the behaviour of the electrons near to the K point is not a direct consequence of the high velocity of electrons in this region. It is a consequence of the linearity of the dispersion relation in this region, not its actual gradient.

Despite this, it is noteworthy that graphene does allow the study of relativistic phenomena (chiral tunnelling, zitterbewegung, Berry's phase, Klein paradox), otherwise not accessible in a tabletop laboratory experiment as a result of the tight-binding Hamiltonian and the Dirac equation being mathematically identical. The interested reader is referred to the many texts on this topic ([31,32] of Chapter 1 and [1,2,9–12]).

3.6   How important is the group velocity at the Fermi level?

What is determined by the group velocity at the Fermi level v_F? Supposing that we apply a voltage across a sample, then (if the sample is a metal or a low-bandgap semiconductor) electrons at or near the Fermi level are free to move into different quantum states in this region such that there will be a net movement of electrons in the direction dictated by the applied voltage. Each electron in this region can move at v_F, hence this velocity (along with the density of mobile electrons) places an upper limit on the conductivity of the sample.

However, for a sample to exhibit a conductivity limited only by v_F and the mobile electron density, the motion of the electrons must be ballistic; that is, once the electron is in a quantum state with a given wavevector, it must stay in that state. If the electrons are regularly scattered into different quantum states (e.g. by defects or by thermal motion of the ions), then their motion is described as diffusive instead. In most materials, this scattering is what limits the conductivity.

Take, for instance, the example of copper. If we look up the calculated dispersion relation near the Fermi level [13] and use Equation 3.15 to calculate the Fermi velocity, we obtain a value of $v_F \approx 2.8 \times 10^5$ ms^{-1}. In tungsten, the Fermi velocity is expected to be $v_F \approx 4$–5×10^5 ms^{-1} [14]. However, the electron drift velocity in metals such as these is ~1 mms^{-1} – in this case a factor of 10^8 lower than the Fermi velocity! This is all due to scattering of electrons from defects and thermal motion of the ions rendering their motion diffusive rather than ballistic. If we wanted to build a miniature electronic circuit using atomically narrow copper wires (say five atomic layers thick), we would expect the drift velocity to decline still further – the extremely large surface area-to-volume ratio would have the same effect as the introduction of many more defects.

In graphene, on the other hand, movement of electrons at ambient temperature is ballistic due to the extremely low concentration of defects [10,15], so the conductivity is limited only by the Fermi velocity. Hence, the high Fermi velocity of $v_F \approx 1.0 \times 10^6$ ms^{-1} is important.

3.7   Vanishing density of states close to the K point in graphene

3.7.1   Density of states in two-dimensional (2D) systems

Readers may be familiar with the concept of the density of quantum states available to electrons in a sample, as a function of their wavevector or energy, in three dimensions. The concept plays an important role in semiconductor physics, for instance, in III–V semiconductors such as gallium arsenide, the large density of states at the direct bandgap between the conduction and valence band leads to applications in photonics.

The density of states in graphene (as a function of energy) displays interesting and unusual behaviour, but first we should discuss the basic mathematics underpinning density of states in two dimensions, because graphene is a genuinely 2D material. Readers requiring a broader view of the subject may also wish to consult standard texts discussing the density of states in three dimensions ([3–5] of Chapter 1).

Consider a particle confined in infinite square well in two dimensions extending from (0,0) to (L,L). The solutions to the Schrödinger equation for this particle are wave functions as follows ([7] of Chapter 2):

$$\psi(k) = A \sin k_x x \sin k_y y \qquad (3.16)$$

We require that $\psi(k) = 0$ at the boundaries of the square well. To ensure this,

$$k_x L = n_x \pi, \quad k_y L = n_y \pi, \quad n_x, n_y \text{ are integers.}$$

The values of k allowed by these conditions form a simple square lattice in reciprocal space (k-space, Figure 3.5). The spacing between adjacent lattice points in k space is (π/L).

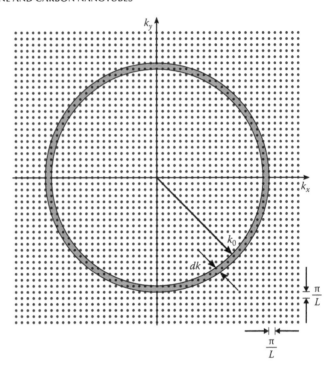

FIGURE 3.5
Quantum states available in two dimensions for a particle with wavevector up to k_0 (area of circle) and states available between k_0 and to $k_0 + dk$ (shaded area).

Hence, the number of lattice points per unit area in k-space is $(L/\pi)^2$, and the number N_{kr} of lattice points per unit area in k-space and per unit area in real space is

$$N_{kr} = \frac{1}{\pi^2} \tag{3.17}$$

If we want to know the number of quantum states available to electrons within any area of k-space, we simply count the number of states in that area. If we want to know the number of quantum states available to electrons within a certain energy interval, we simply use the electronic dispersion relation $E(k)$ to calculate the area in k-space covered by the dispersion relation within that energy interval.

We can derive general expressions for the density of states by considering the number of states in circular areas of k-space in two dimensions. Let us therefore draw a circle in k-space with radius k_0 (Figure 3.5) and count the total number of quantum states $n_t(k_0)$ available within that circle:

$$n_t(k_0) = 2 \times \frac{1}{4} \times \pi k_0^2 \times \left(\frac{L}{\pi}\right)^2 \tag{3.18}$$

Here, we have a factor of 2 to account for the fact that each quantum state may be occupied by two electrons with opposing spin. We then need to divide by 4 to account for the fact that if we replace k_x with $-k_x$ and/or replace k_y with $-k_y$ then the value of $|\psi(k)|^2$ remains the same; that is, it is the same quantum state. Therefore, if we consider more than one quadrant of the circle in Figure 3.5, then we are counting the available quantum states more than once. We then multiply by the area of the circle (πk_0^2) and the number of quantum states available per unit area $((L/\pi)^2)$.

Using Equation 3.18, we can derive an expression for the *density of states* $\rho(k_0)dk$, the number of quantum states available between wavevectors k_0 and $(k_0 + dk)$, per unit area (of real space):

$$\rho(k_0)dk = \frac{1}{L^2}\frac{dn_t}{dk_0}dk = \frac{k_0 dk}{\pi} \qquad (3.19)$$

Note that the density of states per unit area of real space is independent of the dimensions of the sample or other parameters; it depends only on k_0 and the spin of the particles (we would have to divide by 2 to apply it to bosons). We can therefore use this expression for any set of fermions in any area of 2D space.

Frequently, we are interested in the density of states as a function of energy $\rho(E)$ rather than as a function of wavevector. For instance, the large density of states as a function of energy above and below the direct electronic bandgap enables the present optoelectronic applications of III–V semiconductors such as Gallium Arsenide and potential optoelectronic applications of single-walled carbon nanotubes (Chapter 5).

If the dispersion relation $E(k)$ is spherically symmetric (isotropic) about the centre of the circle drawn in Figure 3.5, we can write a simple expression relating $\rho(E)$ to $\rho(k_0)$. Usually, electronic dispersion relations are not isotropic. However, fortuitously, graphene's dispersion relation is isotropic about the most important point; the K point where the conduction and valence bands meet. So we can write that $\rho(E)|dE| = \rho(k_0)|dk|$ and

$$\rho(E)dE = \rho(k_0)\left|\frac{dE}{dk}\right|^{-1}dE \qquad (3.20)$$

3.7.2   Density of states in graphene at the K point

In graphene, according to theory, the density of available quantum states for electrons as a function of energy drops to zero as we approach the K point. We are now in a position to understand this from a mathematical and physical point of view.

From the physical point of view, one may consider the number of states available between energy E and $(E + dE)$ by

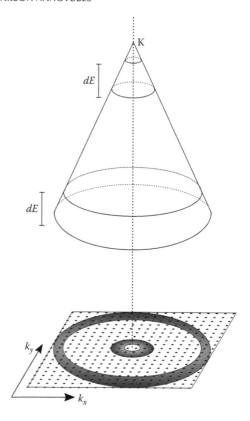

FIGURE 3.6
Density of states in graphene dropping to zero at the K point. Within a given energy interval dE, the section of the Dirac cone closer to the K point maps out a much smaller area in k-space than that within the same energy interval dE further away from the K point. Hence, the number of quantum states available within interval dE drops to zero as the K point is approached.

considering the area of k-space covered by the Dirac cone in this energy interval. One can see (Figure 3.6) that as we get closer to the K point, the area of k-space covered by the Dirac cone in a given energy interval dE gets smaller and smaller, and must eventually approach zero.

From the mathematical point of view, we can consider the density of states as follows. The dispersion relation is spherically symmetric about the K point, so we may use Equation 3.20 to predict the density of states as a function of energy $\rho(E)dE$ as we approach the K point. Setting k_0 as the wavevector leading from the K point to the edge of the Dirac cone ($k_0 = \delta k$), we can combine Equations 3.19 and 3.20 to give

$$\rho(E)dE = \frac{\delta k}{\pi}\left[\frac{dE}{dk}\right]^{-1}dE \tag{3.21}$$

As we move closer to the K point, $k_0 \rightarrow 0$ and hence $\rho(E)dE \rightarrow 0$; in agreement with our conclusion earlier from the simple geometric arguments.

3.8 Cyclotron motion of electrons in graphene

A variety of techniques have been used to study the electronic dispersion relation of graphene. We saw, for instance, ARPES data in Figure 3.4 directly illustrating the linear relationship between E and k. However, the technique first used to verify the linearity of the dispersion relation and speed (group velocity) of the electrons in graphene was the study of the motion and energy levels of the electrons in a magnetic field (cyclotron motion).

3.8.1 Semi-classical model of cyclotron motion

It is possible to understand cyclotron motion of electrons using a simple semi-classical model ([3] of Chapter 1). We can begin by writing down the equation of motion for an electron in a magnetic field, relating the wavevector k and group velocity v to the magnetic field strength B.

$$\hbar \frac{d\mathbf{k}}{dt} = -e\mathbf{v} \times \mathbf{B} \tag{3.22}$$

Here, we are equating the rate of change of momentum of the electron (on the left-hand side) with the Lorentz force on a moving particle in a magnetic field (on the right-hand side). We conclude from inspection of Equation 3.22 that any closed cyclotron orbit of an electron must exclusively consist of motion in a plane perpendicular to B.

For graphene, therefore, the cyclotron motion of the electrons consists of circular orbits around the Dirac cones whilst the magnetic field is applied in the direction perpendicular to the graphene sheet (Figure 3.7).

We continue by writing down an expression for the time taken to traverse the section of the cyclotron orbit marked in Figure 3.7, in terms of the motion of the electron described by its wavevector k.

$$t_2 - t_1 = \int_{t_1}^{t_2} dt = \int_{k_1}^{k_2} \frac{1}{|d\mathbf{k}/dt|} dk \tag{3.23}$$

We can then evaluate $|d\mathbf{k}/dt|$ using Equation 3.22:

$$\left| \frac{d\mathbf{k}}{dt} \right| = \left| -\frac{1}{\hbar} e\mathbf{v} \times \mathbf{B} \right|$$

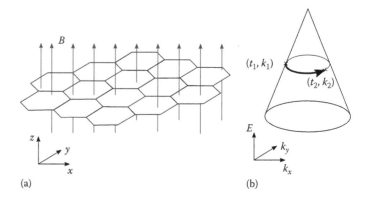

FIGURE 3.7
(a) To study the cyclotron motion of electrons in graphene, we apply a magnetic field normal to graphene lattice. This results in electrons executing cyclotron orbits of constant energy around the Dirac cones (b).

The group velocity of the electrons is defined as follows*:

$$v = \frac{1}{\hbar} \nabla_k E \qquad (3.24)$$

where E is the energy of the electron. Hence,

$$\left| \frac{dk}{dt} \right| = \left| \frac{e}{\hbar^2} \nabla_k E \times B \right|$$

But since $\frac{1}{\hbar} \nabla_k E$ is the group velocity, it is always perpendicular to B. We can hence write

$$\left| \frac{dk}{dt} \right| = \frac{eB}{\hbar^2} \left| \nabla_k E \right|$$

$$t_2 - t_1 = \frac{\hbar^2}{eB} \int_{k_1}^{k_2} \frac{dk}{\left| \nabla_k E \right|} \qquad (3.25)$$

To progress further, we must consider what happens to the orbital period when the energy and wavevector of the orbit

---

* See References 4 and 5 of Chapter 1. ∇_k is simply the ∇ operator in k-space: $\nabla_k = \begin{pmatrix} \dfrac{\partial}{\partial k_x} \\ \dfrac{\partial}{\partial k_y} \end{pmatrix}$.

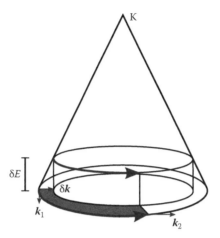

FIGURE 3.8
Schematic of cyclotron orbits near to the K point. The shaded area is the value of the integral in Equation 3.26.

change by small quantities δE and δk as defined in Figure 3.8. We can relate δE and δk using the quantity $\nabla_k E$:

$$\nabla_k E \cdot \delta k = \begin{pmatrix} \dfrac{\partial E}{\partial k_x} \\ \dfrac{\partial E}{\partial k_y} \end{pmatrix} \cdot \begin{pmatrix} \delta k_x \\ \delta k_y \end{pmatrix} = \delta E$$

Applying the ∇_k operator to $E(k)$ gives a result perpendicular to the circular orbit of constant energy and parallel to δk. Hence $\nabla_k E \cdot \delta k = \delta E = |\nabla_k E||\delta k|$ and:

$$t_2 - t_1 = \frac{\hbar^2}{eB} \frac{1}{\delta E} \int_{k_1}^{k_2} \delta k \, dk \tag{3.26}$$

We are interested in studying one complete orbit, the case where $t_2 - t_1 = \tau$, the orbital period. In this case we can show from geometrical considerations (Figure 3.8) that the integral in Equation 3.26 is simply the change in the area A mapped out by the cyclotron orbit due to the change in energy δE. We can therefore write

$$\tau = \frac{\hbar^2}{eB} \frac{\partial A}{\partial E} \tag{3.27}$$

Now that we have an expression for the period τ, we can compare it to the classical expression for the cyclotron frequency $(\omega_c = (2\pi/\tau) = (eB/m_c))$ ([3] of Chapter 1) to obtain an expression for the effective cyclotron mass m_c of the electron:

$$m_c = \frac{\hbar^2}{2\pi} \frac{\partial A}{\partial E} \tag{3.28}$$

3.8.2 Quantization of cyclotron orbits in graphene

According to our treatment of the cyclotron motion so far, the electron can undergo cyclotron motion in a magnetic field with any energy E and any cyclotron mass m_c. But in actual fact the electron can sit only in quantized energy levels separated by energy $\Delta E = \hbar \omega_c$. Using this condition we can state that the allowed energy levels must be given using Equation 3.29. Various proofs of this quantization condition are found in the literature ([2,3,5] of Chapter 1).

$$E_l = \left(l + \frac{1}{2} \right) \frac{\hbar e B}{m_c} \tag{3.29}$$

where l is an integer. The Dirac cones are hence split up into a series of circles (the Landau circles*), as shown in Figure 3.9. As we vary the magnetic field, the circles of allowed states move up and down the Dirac cones.

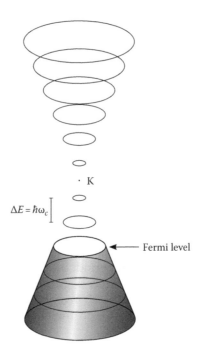

FIGURE 3.9
Landau circles near the K point in graphene: quantization of available electronic energy levels in a magnetic field. Normally the Fermi level lies at the K point, but here we have varied the Fermi level by applying a gate voltage to observe the effect of moving it through successive Landau circles.

---

* Texts dealing with 3D systems refer instead to "Landau tubes".

In a pristine graphene sample, the Fermi level is at the K point. However, the experimental procedure when we study the cyclotron motion of electrons in graphene is to vary the Fermi level by applying a gate voltage ([31,32] of Chapter 1) and (for a given gate voltage and Fermi level) vary the magnetic field. When we do this, we observe large oscillations in carrier density (i.e. density of mobile electrons), with positive spikes in the carrier density when one of the Landau circles exactly intersects the Fermi level. We can work out (and compare to experimental findings) the period (in units of $1/B$) with which these oscillations occur.

According to Equation 3.29, the energy spacing between adjacent Landau circles is

$$\delta E = \frac{\hbar e B}{m_c} = \frac{2\pi e B}{\hbar} \frac{1}{(\partial A/\partial E)} \tag{3.30}$$

Since the Landau circles in graphene (or Landau tubes in typical 3D systems for that matter ([3] of Chapter 1)) are very closely spaced, it is reasonable to state that $(\partial A/\partial E) \approx (\delta A/\delta E)$, where δA and δE are the changes in cross-sectional area and energy between adjacent circles. Hence,

$$\delta A = \frac{2\pi e B}{\hbar}$$

And the total area enclosed by the lth circle is

$$A_l = \left(l + \frac{1}{2}\right) \frac{2\pi e B}{\hbar} \tag{3.31}$$

We can define the area enclosed by the Dirac cone at the Fermi level as A_F, and, neglecting the factor of $1/2$, state that the lth Landau circle will pass through the Fermi level when $A_F = A_l$ or

$$l \times \frac{2\pi e}{\hbar} \times \frac{1}{A_F} = \frac{1}{B}$$

When B is varied (and A_F is kept constant), the interval in B between successive Landau circles passing through the Fermi level is known as the *Fundamental Field* (B_F). It is given as follows:

$$B_F = \frac{1}{\Delta(1/B)} = \frac{\hbar}{2\pi e} A_F \tag{3.32}$$

3.8.3 Cyclotron orbits in graphene: Comparison of theory to experiment

We are now in a position to look at the experimental observations regarding cyclotron orbits in graphene (Figure 3.10) and discover what they tell us about graphene's electronic

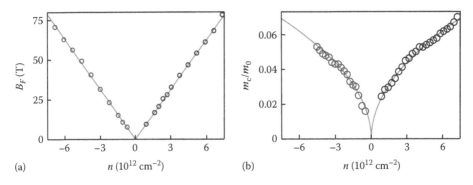

FIGURE 3.10
Experimentally observed fundamental field B_f (a) and electron cyclotron mass m_c (b) in graphene. (Reprinted by permission from Macmillan Publishers Ltd. *Nature*, Novoselov, K.S. et al., 438, 197, Copyright 2005 [Chapter 1].)

dispersion relation. We do this utilizing Equations 3.28 and 3.32, where A_F (which is a function of E) is the parameter which allows us to learn about the dispersion relation. As we vary the Fermi level by applying the gate voltage, $A_F(E)$ changes.

Experimentally ([31,32] of Chapter 1), the key observations (Figure 3.10) are that the electron cyclotron mass m_c is proportional to $\sqrt{n}$ and the *fundamental field* B_F is proportional to n, where n is the mobile carrier density in the graphene sheet:

$$m_c = \varepsilon\sqrt{n} \tag{3.33}$$

$$B_F = \frac{h}{4e}n \tag{3.34}$$

where ε is a constant which is determined experimentally.

Equating expressions (3.32) and (3.34) for B_F and relating A_F to the wavevector of the electron at the Fermi level k_F ($A_F = \pi k_F^2$) gives a relation between the wavevector k_F and the carrier density n:

$$k_F^2 = \pi n \tag{3.35}$$

We then apply Equation 3.28 to the Fermi level to obtain an expression for the cyclotron mass at the Fermi level:

$$m_c = \hbar^2 k \times \frac{\partial k}{\partial E} \tag{3.36}$$

Combining Equations 3.33, 3.35 and 3.36 gives us an expression from which we can obtain the $E(k)$ relation near the Fermi level and calculate the group velocity of the carriers:

$$\frac{\partial k}{\partial E} = \frac{\varepsilon}{\hbar^2\sqrt{\pi}} \tag{3.37}$$

Thus the experimental data contained in Figure 3.10 directly prove the key feature of the electronic dispersion relation for graphene: That close to the K point the electronic dispersion relation is linear ($(\partial k/\partial E)$ is a constant). The experimentally determined value of ε ([31,32] of Chapter 1) allows us to experimentally determine a value for the group velocity of electrons in graphene at the Fermi level (using Equation 3.15): $v_F \approx 1.0 \times 10^6$ ms^{-1}, in agreement also with the findings of ARPES experiments on graphene.

References

1. AH Castro Neto, F Guinea, NMR Peres, KS Novoselov and AK Geim, *Rev. Mod. Phys.* **81**, 109 (2009).

2. E McCann, Chapter 8: Electronic properties of monolayer and bilayer graphene, in *Graphene Nanoelectronics*, H Raza (ed.), Springer, Berlin (2012).

3. M Koshino and T Ando, Chapter 8: Electronic properties of monolayer and bilayer graphene, in *Physics of Graphene*, H Aoki and MS Dresselhaus (eds.), Springer, Berlin (2014).

4. O Vilkov et al., *Sci. Rep.* **3**, 2168 (2013).

5. I Pletikosić, M Kralj, P Pervan, R Brako, J Coraux, AT N'Diaye, C Busse and T Michely, *Phys. Rev. Lett.* **102**, 056808 (2009).

6. A Bostwick et al., *Science* **328**, 999 (2010).

7. VB Berestetsky, EM Lifshitz and LP Pitaevsky, *Relativistic Quantum Theory*, Pergamon Press, Oxford (1971).

8. T Ohlsson, *Relativistic Quantum Physics*, CUP, Cambridge (2011).

9. MI Katsnelson, *Graphene: Carbon in 2 Dimensions*, CUP, Cambridge (2012).

10. AK Geim and KS Novoselov, *Nat. Mater.* **6**, 183 (2007).

11. MI Katsnelson, KS Novoselov and AK Geim, *Nat. Phys.* **2**, 620 (2006).

12. TN Rusin and W Zawadzki, *Phys. Rev. B* **78**, 125419 (2008).

13. GA Burdick, *Phys. Rev.* **129**, 138 (1963).

14. J Kollár, *Solid State Commun.* **27**, 1313 (1978).

15. AK Geim, *Science* **324**, 1530 (2009).

4

Advanced Considerations on the Electronic Dispersion Relation of Graphene

Study of the cyclotron motion of electrons in graphene provided us with key information about the electronic dispersion relation in the material: a proof that the dispersion relation close to the K point is linear and a measurement of the group velocity of electrons near to the K point. We can, however, learn still more about the dispersion relation by studying the behaviour in larger magnetic fields, the quantum Hall effect (QHE) in graphene.

It is through study of the QHE that we can directly observe the requirement to treat the electrons in graphene using the Dirac equation, as massless relativistic particles. We will begin with a general description of the classical and QHEs, before moving on to the anomalous QHE in graphene. Readers already familiar with the classical Hall effect may therefore wish to skip to Section 4.1.2, and readers familiar with the QHE may wish to skip to Section 4.2.

4.1 The Hall effect

4.1.1 A classical treatment of the Hall effect

The Hall effect [1] is an effect observed when an electric current is put through a conductor or semiconductor under the influence of a magnetic field perpendicular to the current: A voltage is generated in the direction perpendicular to both the magnetic field and the current. The Hall effect originally observed in 1879 in metals [1] can be understood in simple classical terms. Under the influence of the magnetic field in the z direction (Figure 4.1), the electrons forming the current in the x direction are deflected in the y direction due to the Lorentz force:

$$F = -ev \times B$$

An accumulation of electrons therefore forms on one side of the sample, generating the Hall voltage V_H. Due to the Hall voltage, an electric field E_y (equal to (V_H/d)) is generated in the y direction which acts to prevent further deflection of the electrons. Hence, the total force on the moving electron is given by

$$F = -e\left(E_y \hat{y} + v \times B\right)$$

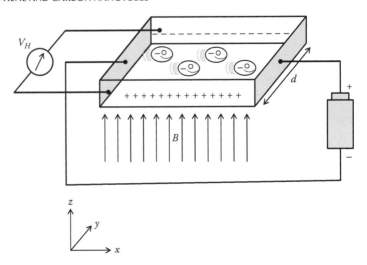

FIGURE 4.1
Hall effect experimental geometry. Under the influence of the applied voltage mobile electrons drift to one side of the sample (in the positive y direction) due to the Lorentz force from the magnetic field B (applied in the positive z direction). This process occurs until, in the steady state, a concentration of charge has built up on one side of the sample to provide a Coulomb force to counteract the Lorentz force. The voltage across the sample in the y direction (the Hall voltage V_H) can be measured to obtain a value for the electric field strength E_y in this direction.

In the steady state, we observe that the motion of the electrons will be in the x direction and set $F = 0$ to give

$$eE_y = ev_x B \qquad (4.1)$$

Substituting the current density in the x direction ($J_x = -N_e e v_x$) into Equation 4.1, we can obtain an expression for a constant called the Hall coefficient, R_H. This can be calculated from the experimental data. From R_H we can obtain the density N_e of mobile electrons in the metal:

$$R_H = \frac{E_y}{J_x B} = -\frac{1}{N_e e} \qquad (4.2)$$

The Hall effect provided clear evidence that the charge carriers in metals were electrons instead of protons and later that in certain semiconductors the principal charge carriers were holes (vacant quantum states in the valence band) instead of electrons in the conduction band.

We can also define expressions for different components of the conductivity tensor σ, which we will utilize in our treatment of the quantum Hall effect (QHE) in the next section. We begin by writing down an expression for the components of the current density J in the x and y directions, in terms of the conductivity tensor $\boldsymbol{\sigma}$ and electric field components:

$$J_x = \sigma_{xx}E_x + \sigma_{xy}E_y \tag{4.3}$$

$$J_y = \sigma_{yy}E_y + \sigma_{yx}E_x = \sigma_{xx}E_y - \sigma_{xy}E_x \tag{4.4}$$

We have assumed that the sample is isotropic in our manipulation of Equation 4.4; that is, we can write $\sigma_{xx} = \sigma_{yy}$ and $\sigma_{yx} = -\sigma_{xy}$. In the steady state in a Hall effect experiment however, $J_y = 0$ so using Equation 4.4 we can write

$$\frac{E_y}{E_x} = \frac{\sigma_{xy}}{\sigma_{xx}} \tag{4.5}$$

We can then define the relevant components of the resistivity tensor $\boldsymbol{\rho}$ in the steady state:

$$\rho_{xx} = \frac{E_x}{J_x} = \frac{E_x}{\sigma_{xx}E_x + \sigma_{xy}E_y} = \frac{\sigma_{xx}}{\sigma_{xx}^2 + \sigma_{xy}^2} \tag{4.6}$$

$$\rho_{xy} = \frac{E_y}{J_x} = \frac{\sigma_{xy}}{\sigma_{xx}^2 + \sigma_{xy}^2} \tag{4.7}$$

We can also define ρ_{xy} using the Hall coefficient (Equation 4.2):

$$\rho_{xy} = \frac{E_y}{J_x} = -\frac{B}{N_e e} \tag{4.8}$$

Note that, whilst the conductivity tensor $\boldsymbol{\sigma}$ was defined as the generic conductivity tensor for that material (before we had specified any experimental conditions or constraints), the resistivity tensor $\boldsymbol{\rho}$ in Equations 4.6 and 4.7 has been defined for the specific case of $J_y = 0$. Also, whilst we have defined ρ_{xx} using $\rho_{xx} = (E_x/J_x)$ (Equation 4.6), we have not defined σ_{xx} using $\sigma_{xx} = (J_x/E_x)$. Instead we wrote Equation 4.3. Hence, we cannot simply write $\rho_{ij} = (1/\sigma_{ij})$ for each separate component of the two tensors $\boldsymbol{\sigma}$ and $\boldsymbol{\rho}$.

4.1.2   Quantum Hall effect

Thus far, we have considered electrons as classical particles; their energy, mobility, etc., can take any value. However, when we study the Hall effect in systems where the electrons are confined to a 2D space,* the effect observed (the quantum Hall effect or QHE) can only be understood by treating the electrons as particles with quantized energy levels. For instance, in a classical Hall effect experiment, the resistivity in the x direction remains constant, as expected. If the electrons are

---

* In this context the system counts as 2D if, in the z direction, the electrons are confined in a space sufficiently small to cause the electrons to sit in well-separated quantized energy levels due to their confinement in this axis. So the sample need not be quite as thin as graphene: A thickness of ≈ 20 nm is adequate [2].

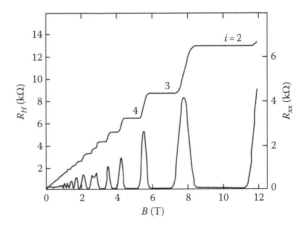

FIGURE 4.2

Transverse (ρ_{xy}) and longitudinal (ρ_{xx}) resistivity components of a gallium arsenide–based heterostructure device subjected to an increasing magnetic field. The transverse (longitudinal) resistivity shows steps (spikes) at certain quantized values. From a classical treatment of the Hall effect, we would expect ρ_{xy} to increase linearly and ρ_{xx} to be a constant. (Reprinted from Gallop, J., *Philos. Trans. R. Soc. A*, 363, 2221, 2005. With permission of the Royal Society.)

classical particles passing through the sample undeviated by the equal and opposite forces due to the electric and magnetic fields in the y direction, then the scattering due to defects, impurities, etc., which causes a finite resistivity should be the same regardless of the magnetic field strength. In a QHE experiment, however, the resistivity (in the x and y directions) changes instead in quantized steps (Figure 4.2).

To understand this, we must clearly consider the electron under the influence of the magnetic field to sit in quantized energy levels. These are the Landau cyclotron energy levels discussed in the previous chapter (Equation 3.29). In Chapter 3, we discussed the cyclotron motion of electrons in graphene in very small magnetic fields, leading to very closely spaced energy levels. To observe the QHE we apply much larger magnetic fields, leading to more widely spaced energy levels. Assuming that the electrons are sitting in the lowest energy level of the quantum well formed due to their confinement in the z direction, we may begin by recalling Equation 3.29:

$$E_n = \left(n + \frac{1}{2} \right) \frac{\hbar e B}{m_c} \qquad (4.9)^*$$

---

* In full, we should write $E_n = \left((n + (1/2))(\hbar e B / m_c) \pm \mu_B B \right)$ to take account of the two possible orientations of the electron spin with respect to the magnetic field (μ_B is the Bohr magneton). Usually, however, this splitting is too small to be observed experimentally. Equating $\mu_B B \approx (3/2) k_B T$ we can see that, for $B = 10\,T$, thermal excitation would prevent this splitting from being observed for $T \approx 4\,K$.

Therefore, the (otherwise continuous) density of states in the system as a function of energy is split into a series of discrete energy levels as shown in Figure 4.3.

To explain the observations in Figure 4.2 in a quantitative manner, however, we need to calculate how many quantum states are available for electrons in each Landau level. Just as in our treatment of cyclotron orbits in much weaker magnetic fields in Chapter 3, we expect that the total number of quantum states available for electrons in a given energy interval is not changed by the application of the magnetic field. We can use this fact to calculate the number of states available per Landau level. We start by equating the energy of the nth Landau level with the wavevector k_n which would correspond to in the absence of the field:

$$E_n = \left(n + \frac{1}{2}\right)\frac{\hbar eB}{m_c} = \frac{\hbar^2 k_n^2}{2m^*} \tag{4.10}$$

We then equate the cyclotron mass m_c with the effective mass m^* and obtain an expression for the wavevector k_n of the nth Landau level in magnetic field B:

$$k_n^2 = 2\left(n + \frac{1}{2}\right)\frac{eB}{\hbar} \tag{4.11}$$

We can use this expression, combined with our expression for the density of states per unit area of k-space and per unit area of real space ($N_{kr} = (1/\pi^2)$, Equation 3.14), to calculate the number of states available in the energy/wavevector interval occupied by the nth Landau level, N_n:

$$N_n = \left(\pi k_{n+1}^2 - \pi k_n^2\right) \times \frac{1}{\pi^2} \times \frac{1}{4} \times 2 = \frac{k_{n+1}^2 - k_n^2}{2\pi} \tag{4.12}$$

Just as in the previous chapter, we have taken the area of k-space in the relevant wavevector interval, multiplied by a factor of 1/4 to account for the fact that we can only count the quantum states from one quadrant of the circle, then by a factor of 2 to account for spin. Substituting for k_n and k_{n+1} using Equation 4.11 gives

$$N_n = \frac{2eB}{h} \tag{4.13}$$

We can hence calculate the total electron density N_e when j Landau levels are completely filled:

$$N_e = j \times \frac{2eB}{h} \tag{4.14}$$

We expect the longitudinal conductivity σ_{xx} to be small when the condition (4.14) is met for integer j, because this is the case

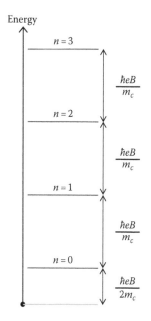

Energy

$n = 3$

$\frac{\hbar eB}{m_c}$

$n = 2$

$\frac{\hbar eB}{m_c}$

$n = 1$

$\frac{\hbar eB}{m_c}$

$n = 0$

$\frac{\hbar eB}{2m_c}$

FIGURE 4.3
Energy levels for electrons confined in a 2D sample in a magnetic field. The number of quantum states available in each energy level is that which ensures that the total number of quantum states available for the electrons up to a given energy is the same as in the absence of the magnetic field.

when the chemical potential lies exactly equidistant from the jth and $(j+1)$th Landau levels; the jth level is full and the $(j+1)$th level is empty. This is effectively the case where the application of the magnetic field has turned the material into an insulator. Note (from Equation 4.6) that, when σ_{xx} tends to zero, we also expect the longitudinal resistivity ρ_{xx} to tend to zero! We can resolve this apparent contradiction by recalling that we have defined the conductivity $\boldsymbol{\sigma}$ and resistivity $\boldsymbol{\rho}$ tensors for different experimental conditions.

On the other hand, we expect σ_{xx} to be large when a Landau level is exactly half-full. In this case there will be many vacant quantum states available for the carriers to occupy to produce a net movement of carriers (current) in the x direction. Analogous to Equation 4.14, the condition for large σ_{xx} is

$$N_e = \left(j + \frac{1}{2}\right) \times \frac{2eB}{h} \tag{4.15}$$

We can observe the spikes in σ_{xx} by varying the magnetic field or by varying N_e (by adjusting the gate voltage). In the case of the data in Figure 4.2, we vary the field and hence observe spikes in ρ_{xx} with period in B of

$$\Delta\left(\frac{1}{B}\right) = \frac{2e}{hN_e} \tag{4.16}$$

We can also predict exactly the quantized values of ρ_{xy} shown in Figure 4.2. Recalling Equation 4.8, we can substitute the values obtained for N_e in Equation 4.14 to obtain

$$\left|\rho_{xy}\right| = \frac{1}{j} \times \frac{h}{2e^2} \tag{4.17}$$

When there are no partially filled Landau levels, $\sigma_{xx} = 0$ and the relation between ρ_{xy} and σ_{xy} (Equation 4.7) reduces to the simple formula: $\rho_{xy} = (1/\sigma_{xy})$. Hence,

$$\sigma_{xy} = j \times \frac{2e^2}{h} \tag{4.18}$$

We can see that the quantized values of ρ_{xy} and σ_{xy} are determined only by the values of Planck's constant h and the electronic charge e! We can plot the expected variation in ρ_{xx} and ρ_{xy} as a function of B at constant N_e, or as a function of N_e (varied by changing the gate voltage) at constant B (Figure 4.4).

It is important to understand at this stage that the energy gap between adjacent Landau levels such as those leading to the data shown in Figure 4.2 is extremely small. The sample needs to have vanishingly small concentrations of defects and impurities; simply due to the fact that collisions with defects and impurities reduce the time the electron spends in a single quantum state and hence leads to uncertainty in its energy due to the energy–time uncertainty principle $\Delta E \Delta t \geq (\hbar/2)$.

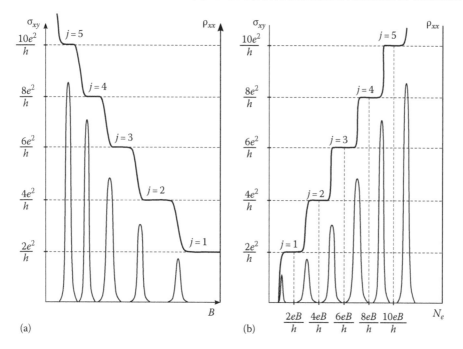

FIGURE 4.4

Schematics showing variation in σ_{xy} and ρ_{xx} as a function of B at constant N_e (a) and as a function of N_e at constant B (b). In both cases we observe plateaux in σ_{xy} at quantized values given by Equation 4.18, the jth plateau corresponding to the jth Landau level being full. The spikes in ρ_{xx} occur each time a new Landau level crosses the Fermi level, at values of N_e given by Equation 4.15.

Even this small uncertainty in energy is enough to cause the Landau levels to merge into each other.

In most cases, the combination of this broadening of the Landau levels and thermal energy prevents the observation of the QHE at temperatures above ~30 K. The splitting in the Landau levels due to spin is even harder to observe – in the data shown in Figure 4.2, for instance, it is not visible.

4.1.3   What can we learn from the quantum Hall effect?

In general terms, the quantization of the Hall effect occurs at resistivity given in multiples of h/e^2 so provides a means to define the unit of resistance ([2] of Chapter 1). As far as the specific sample is concerned, the observation of the QHE provides information about the crystalline quality of the sample and the density of states as a function of energy. We discussed earlier the requirement to observe the QHE – that the sample should have very few defects. The information provided about the density of states stems from the fact that, to observe the periodic changes in resistivity shown in Figure 4.2, there must be quantum states available (in the zero magnetic field case) from which the Landau levels shown in Figure 4.3 can be formed.

4.2   Quantum Hall effect in graphene

The QHE in graphene was first observed in 2005, by two groups publishing independently in Nature ([31,32] of Chapter 1). The observation of the QHE in graphene is remarkable for several reasons. First, the fact that it has been observed at ambient temperature ([33] of Chapter 1), the first time the effect has been observed at temperatures higher than 30 K [4,5]. Second, the nature of the quantization observed is different to both the conventional QHE (Section 4.1.2) and the fractional QHE ([3] of Chapter 1). This is a direct consequence of the need to treat the electrons in graphene as massless relativistic particles. Third, a spike in ρ_{xx} is observed from a Landau level at the K point, despite the expectation (Section 3.7.2) that the density of states available for electrons drops to zero at this point.

4.2.1   Quantization of the Hall effect in graphene

The first remarkable point to make concerning the quantization of the Hall effect in graphene is the fact that the Hall conductivity σ_{xy} is quantized (Figure 4.5) with plateaux at half-integer filling factors (as opposed to integer filling factors for the conventional QHE).

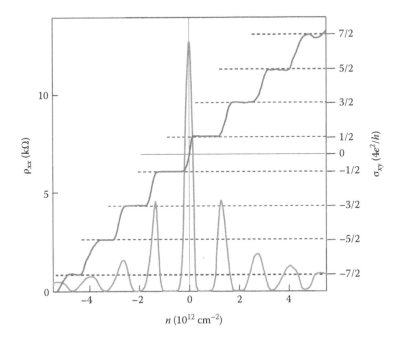

FIGURE 4.5

The longitudinal resistivity ρ_{xx} (spikes, left hand axis) and Hall conductivity σ_{xy} (plateaux, right hand axis) are plotted as a function of mobile carrier concentration n in a pristine graphene sample. (Reprinted by permission from Macmillan Publishers Ltd. *Nature*, Novoselov, K.S. et al., 438, 197, Copyright 2005.)

The plateaux of σ_{xy} at integer filling factors (Equation 4.18 and Figure 4.4) in the conventional QHE is a direct consequence of the energy levels quantized in half-integer multiples that result from treating the electrons as particles with mass (Equations 4.9 and 4.10). Thus, the only explanation for the plateaux of σ_{xy} not occurring at integer filling factors when the QHE is observed in graphene is that the Landau energy levels cannot be obtained by treating the electrons in this way.

As we can see from our discussion in the previous chapter, in the vicinity of the chemical potential and Fermi level at the K point the electrons behave as massless particles; the relationship between their energy and their wave-vector/momentum is linear rather than quadratic. As a result, the approximation to the tight-binding Hamiltonian matrix operator which is valid for the electrons close to the K point is Equation 4.19, which is mathematically identical to the Dirac equation. The Dirac equation is an equivalent to the Schrödinger equation, but is consistent with special relativity (i.e. it is relativistically invariant). So, unlike the Schrödinger equation it can be used to treat particles moving at relativistic velocities. The Dirac equation applies specifically to spin-1/2 fermions.

$$v_F \begin{pmatrix} 0 & ip_x \pm p_y \\ -ip_x \pm p_y & 0 \end{pmatrix} C_i = E_i(k) C_i \qquad (4.19)$$

Here, p_x and p_y are the components of the electron's momentum, C_i is a two-component matrix describing the coefficients with which the atomic electron wave functions combine to form the tight-binding wave function, the subscript i denotes that we are obtaining the energies of either the electrons in the conduction band or the holes in the valence band, and the $\pm$ sign indicates which of the two inequivalent K points we are considering.

Readers wishing to acquaint themselves with the mathematics in more detail are referred to the derivation of this equation in Appendix B (resulting in Equation B.33, from which Equation 4.19 is reproduced), and to the relevant literature ([1,2,7,8] of Chapter 3). Here we will note the key points regarding the application of the Dirac equation to graphene, points which are self-evident from the mathematics:

1. Where the speed of light c appears in the Dirac equation, the group velocity v_F of electrons near the K point appears in Equation 4.19. This is a result of the mathematical equivalence between the two equations and not of the value of v_F. The fact that v_F is a significant fraction of the speed of light is not relevant; it is a coincidence that v_F is (just about) a relativistic

velocity and that we use the relativistic Dirac equation to describe the motion of the electrons.

2. Dirac found the existence of spin necessary to reconcile quantum mechanics and special relativity; thus the possible orientations of the particle spin appear as a variable in the Dirac equation. However, in the mathematically equivalent Equation 4.19, the same variable is accounting not for the two possible orientations of the electron spin but for the two inequivalent* Dirac cones and K points in the reciprocal lattice where there are quantum states available for the electron.

3. In Dirac's relativistic quantum mechanics, the particle wave function consists of four components. These are to account for the two possible orientations of the particle spin as discussed earlier and also to allow for the particle to be a particle or an antiparticle. In Equation 4.19, instead of accounting for the possibility of an antiparticle, we account for holes as well as electrons.

We can note from conditions 2 and 3 that, although the electrons have a spin of 1/2, this variable does not appear in the tight-binding Hamiltonian matrix operator.

The energy eigenvalues obtained by solving the Dirac equation for a massless particle in a magnetic field are ([31] of Chapter 1 and [7] of Chapter 3)

$$E_{n\pm} = c\sqrt{2e\hbar B\left(n + \frac{1}{2} \pm \frac{1}{2}\right)} \qquad (4.20)$$

where n is the quantum number associated with the energy level and the $\pm(1/2)$ term accounts for the two possible orientations of the spin. Replacing the speed of light c with the Fermi velocity v_F gives us the expression for the graphene Landau levels in a magnetic field:

$$E_{n\pm} = v_F\sqrt{2e\hbar B\left(n + \frac{1}{2} \pm \frac{1}{2}\right)} \qquad (4.21)$$

where the $\pm(1/2)$ term now accounts for the existence of two inequivalent K points as discussed earlier, instead of the possible orientations of the spin. This property is hence referred to as "pseudospin".

We can therefore plot an energy level diagram showing the various low-lying energy levels $E_{n\pm}$ (Figure 4.6). At each energy where Landau levels exist, we must count the quantum states available in the two different degenerate levels

---

* Each of the six vertices of the first Brillouin zone touches a different K point, but most can be joined by reciprocal lattice vectors so we are left with only two K points which are not equivalent.

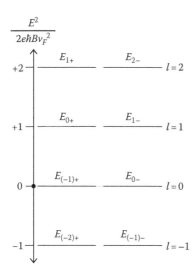

FIGURE 4.6
Energy level diagram of low-lying Landau levels in graphene. The quantum number l is used to describe each set of two degenerate energy levels. l refers to the $E_{(l-1)+}$ energy level and the E_{l-} energy level. Note that the vertical axis is proportional to E^2.

at that energy: For any n, the energy levels E_{n+} and $E_{(n+1)-}$ are degenerate. We therefore introduce the quantum number l, which is used to describe each set of two degenerate energy levels. l refers to the $E_{(l-1)+}$ energy level and the E_{l-} energy level.

We can, just as in our treatment of the conventional QHE earlier, count the number of quantum states available to the lth energy level by counting the number of states N_l which would exist in that energy interval in the absence of a magnetic field. We have to count separately the states contributing to the two levels at this energy:

$$N_l = \frac{1}{\pi^2} \times 2 \times \frac{1}{4} \times \left[\left(\pi k_{(l+1)-}^2 - \pi k_{l-}^2 \right) + \left(\pi k_{l+}^2 - \pi k_{(l-1)+}^2 \right) \right]$$

Here, just as in our treatment earlier of the conventional QHE, we multiply by 2 to account for spin and divide by 4 to avoid quadruple – counting the number of quantum states available. The two different Landau levels we have labelled with l both have the same wavevector.* Hence our expression for N_l simplifies to

$$N_l = \frac{1}{\pi} \left(k_{(l+1)-}^2 - k_{l-}^2 \right) \tag{4.22}$$

---

* This is because we are, in both cases, measuring the wavevector relative to the K point and the dispersion relation around each of the two K points we account for is the same.

k can be evaluated by equating the expression for the Landau level (Equation 4.21) with the linear dispersion relation between E and k:

$$E_{n\pm} = v_F \sqrt{2e\hbar B \left(n + \frac{1}{2} \pm \frac{1}{2} \right)} = \hbar k_{n\pm} v_F \qquad (4.23)$$

Hence,

$$k_{l-}^2 = \frac{2eB}{\hbar} l$$

giving

$$N_l = \frac{4eB}{h} \quad \text{for } l \neq 0 \qquad (4.24)$$

Why is this expression not valid for $l = 0$? It is because the $l = 0$ level consists of the E_{0-} Landau level for electrons, and the $E_{(-1)+}$ energy level, which is the uppermost energy level for holes. So as far as the electron density N_e is concerned, the $l = 0$ level has half the degeneracy of all the other levels. Therefore, if we have energy levels from 0 to l completely filled, the total electron density N_e will be

$$N_e = \left(l + \frac{1}{2} \right) \times \frac{4eB}{h} \qquad (4.25)$$

We can therefore (correctly) predict that the spikes in ρ_{xx} will occur whenever a new Landau level is being filled, that is, when

$$N_e = l \times \frac{4eB}{h} \qquad (4.26)$$

In addition, we can exactly predict the quantized values of the transverse conductivity σ_{xy} as shown in Figure 4.5. The successive quantized values of σ_{xy} correspond to successive sets of two degenerate energy levels being full (Figure 4.6) and to successive quantized values of the electron density as given by Equation 4.25. At each plateau, $\sigma_{xx} = 0$ (in this and other QHE experiments, the experimentally measured value of σ_{xx} does drop to zero within experimental error). Hence, Equation 4.7 simplifies to $\sigma_{xy} = (1/\rho_{xy})$ and (using Equations 4.8 and 4.25) we can write

$$|\sigma_{xy}| = \frac{N_e e}{B} = \left(l + \frac{1}{2} \right) \times \frac{4e^2}{h} \qquad (4.27)$$

Here, we have correctly predicted the experimentally observed quantized values of the transverse conductivity σ_{xy} for graphene (Figure 4.5). These depend only on the values of the fundamental constants e and h, but are quantized in a different manner to the case of the conventional QHE. To achieve this, it has been necessary to treat the electrons near the K point in graphene using a Hamiltonian matrix operator identical to that in the Dirac equation, the relativistic equivalent of the Schrödinger equation. Study of the QHE in graphene therefore provides key proof of the remarkable nature of graphene's electronic dispersion relation.

4.2.2   Density of states at the K point in graphene

We have seen earlier that our theoretical understanding of graphene's electronic dispersion relation allows us to correctly predict certain observations when studying the QHE in graphene, such as the quantized values of σ_{xy} (Figure 4.5). There is, however, one important experimental observation in Figure 4.5 which cannot be reconciled directly with our theoretical understanding of graphene's electronic dispersion relation. We see a peak in ρ_{xx} and a shift in σ_{xy} at $N_e = 0$, proving that a Landau level (the E_{0-} level) is being filled at this point. Since the application of the magnetic field does not change the total number of quantum states available in a given energy interval, the fact that this Landau level actually exists suggests that there are quantum states available close to $E = 0$, even in the absence of a magnetic field. This is, at first sight, at odds with our theoretical prediction in Section 3.6.2 that the density of states drops to zero at the K point.

How can this (apparent) discrepancy be explained? There are several possibilities.

First, we shall see in Chapter 8 that all real graphene sheets contain ripples on a microscopic scale and in Chapter 11 that this is necessary to make graphene thermodynamically stable. Hence, the completely flat graphene sheet which we assume when we perform the tight-binding calculation in Chapter 3 does not exist and the prediction of zero density of states at the K point is a result of a tight-binding calculation which does not quite reflect the true nature of graphene.

Second, there may be an explanation in terms of quantization of conductivity (even in the case of zero magnetic field). Analogous to zero-point energy, there may be some minimum value beneath which the conductivity cannot fall and to achieve this always some quantum states available for electrons to occupy ([31] of Chapter 1 and [10] of Chapter 3). This can be understood as a requirement for the mean free path to always be longer than the de Broglie wavelength of the electrons near the Fermi level ([31] of Chapter 1).

Third, it is of course not possible to conduct a QHE experiment at the ideal temperature of absolute zero. Even at liquid helium temperatures (the data shown in Figure 4.5 were collected at $T = 4$ K), there is some thermal energy available. In addition, the Landau levels may be broadened slightly due to the presence of defects and impurities as discussed earlier in general terms. The combination of these factors could cause a number of quantum states to be available for the E_{0-} Landau level.

4.2.3   Observation of the quantum Hall effect at ambient temperature in graphene

In contrast to previous observations of the QHE, which required temperatures below 30 K [4,5], in graphene the effect has been observed at ambient temperature (300 K) ([33] of Chapter 1). Undoubtedly, this is partially due to the exceptionally high crystalline quality of the graphene samples available in the laboratory. However, it is also another demonstration of the fact that the electron energy levels in the magnetic field are determined by the Dirac-like equation; the Dirac equation applied to graphene (Equation B.33) gives energy levels (Equation 4.21) which are much further apart than those typically observed in the conventional QHE (Equation 4.9).

To demonstrate this, let us begin by finding the gap between the E_{2-} and E_{1-} Landau levels in graphene predicted by Equation 4.21 (using $B = 10T$):

$$E_{2-} - E_{1-} = v_F \sqrt{2e\hbar B}\left(\sqrt{2}-1\right) = 0.05 \text{ eV} \tag{4.28}$$

Using the Landau energy levels in the conventional QHE, however (Equation 4.9), we find the gap between adjacent energy levels as

$$E_{n+1} - E_n = \frac{\hbar e B}{m_c} = 0.001 \text{ eV} \tag{4.29}$$

where, to obtain the numerical value, we have assumed that the cyclotron mass m_c is the same as the mass of the free electron. Generally they are of the same order of magnitude [6]. In both cases we have assumed $B = 10\,T$, which is typical of the magnetic field strength utilized in QHE experiments.

References

1. EH Hall, *Am. J. Math.* **2**, 287 (1879).

2. R Dingle, W Wiegmann and CH Henry, *Phys. Rev. Lett.* **33**, 827 (1974).

3. J Gallop, *Philos. Trans. Royal Soc. A*, **363**, 2221 (2005).

4. SQ Murphy et al., *Physica E (Amsterdam)* **6**, 293 (2000).

5. G Landwehr et al., *Physica E (Amsterdam)* **6**, 713 (2000).

6. G Dresselhaus, AF Kip and C Kittel, *Phys. Rev.* **98**, 368 (1955).

5 Electronic Dispersion Relation of Single-Walled Carbon Nanotubes (SWCNTs)

5.1 Some introductory notes

In this chapter, we successfully model the electronic dispersion relations of single-walled carbon nanotubes (SWCNTs). Whilst we use the 2D electronic dispersion relation for graphene as a starting point, the dispersion relation for SWCNTs is different in a fundamental way because it is one-dimensional (1D). The component of the electron wavevector in the axial direction (along the SWCNT) can take any value; SWCNTs which are synthesized in the laboratory are far too long for quantum confinement of electrons in the axial direction to play any role. However, in the circumferential direction (i.e. the direction circumnavigating the SWCNT), the electron wavevector can only take certain discrete values, those values for which the SWCNT circumference is an integer multiple of the electron wavelength, hence leading to the electron wave function being continuous.

We may therefore model the electronic dispersion relation of an SWCNT by taking the 2D electronic dispersion relation for graphene (Figure 3.2) and calculating the allowed wavevectors for the electron in the SWCNT – they are a series of 1D lines. Where these lines sit in the 2D graphene dispersion relation depends on the diameter and chirality of the SWCNT, the information encapsulated in the (n,m) indices which we defined in Section 1.4.

If one of these lines intersects with the K point where graphene's conduction and valence bands meet, the SWCNT is a zero-bandgap semiconductor (referred to as a metallic nanotube); otherwise, it is a semiconducting nanotube. So by considering this quantization, we can understand why the electronic dispersion relation for an SWCNT varies enormously depending on the diameter and specific (n,m) indices of the tube. In contrast to graphene, SWCNTs usually have a large density of states at the Fermi level. This can also be modelled using the simple approach outlined here. This large density of states, combined with the non-zero electronic bandgap, leads to potential applications in optoelectronics.

The procedure outlined here is called zone-folding, as it is analogous to folding up the 2D first Brillouin zone of graphene into a series of 1D lines forming the first Brillouin zone and other permitted wavevectors for the electron in the SWCNT. The zone-folding procedure is remarkably successful in the prediction of properties such as the metallic or semiconducting nature of SWCNTs using their (n,m) assignment, but also has limitations as discussed in later sections.

5.2   Primitive unit cell and first Brillouin zone of SWCNTs

5.2.1   Primitive unit cell

Whilst the primitive unit cell of graphene contains just two atoms, the primitive unit cell of the SWCNT is much larger. The smallest unit from which we can reproduce the structure of an SWCNT purely by translation is a unit cell which completely circumnavigates the tube and also extends some distance along the tube. In the circumferential direction, the SWCNT lattice vector is simply the chiral vector C_h, which we define as integer (n,m) multiples of the graphene lattice vectors to ensure that the SWCNT lattice is continuous (Section 1.4):

$$C_h = na_1 + ma_2 \tag{5.1}$$

The other SWCNT lattice vector, in the axial direction, is called the translation vector T. It is the shortest vector which runs perpendicular to C_h (i.e. parallel to the SWCNT axis) and which begins and ends on a lattice point (i.e. is some integer multiple of the graphene lattice vectors). The integers (n,m) completely define the structure of the SWCNT. Hence, just like C_h, T can be written in terms of (n,m) and the graphene lattice vectors a_1 and a_2:

$$T = t_1 a_1 + t_2 a_2 \tag{5.2}$$

where t_1 and t_2 are integers. Using the expression for a_1 and a_2 (Equation 1.1) and the condition $C_h \cdot T = 0$, we can obtain expressions for t_1 and t_2 (Equation 5.3). p is the greatest common divisor of $(2m + n)$ and $(2n + m)$. Figure 5.1 shows the chiral (C_h) and translation (T) vectors for a (10,5) SWCNT:

$$t_1 = \frac{2m + n}{p} \tag{5.3}$$

$$t_2 = -\frac{2n + m}{p}$$

5.2.2   First Brillouin zone and quantization of the electron wavevector

We are now ready to define the reciprocal lattice vectors of the SWCNT, which determine the dimensions of the first Brillouin zone. Since the primitive unit cell of the SWCNT in real space is much larger than that of graphene, we expect the primitive unit cell in reciprocal space (the first Brillouin zone)

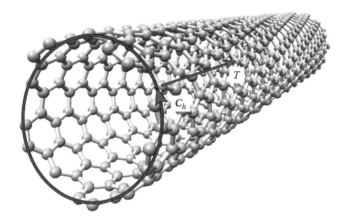

FIGURE 5.1
Primitive unit cell of a (10,5) SWCNT, defined by the chiral vector (C_h) and translation vector (T).

to be much smaller. We begin by defining the reciprocal lattice vectors K_1 and K_2 in the normal manner:

$$C_h \cdot K_1 = 2\pi \qquad (5.4)$$

$$C_h \cdot K_2 = 0$$

$$T \cdot K_2 = 2\pi$$

$$T \cdot K_1 = 0$$

We can write the SWCNT reciprocal lattice vectors in terms of the graphene reciprocal lattice vectors, $K_1 = \alpha_1 b_1 + \beta_1 b_2$ and $K_2 = \alpha_2 b_1 + \beta_2 b_2$, and (making no prior assumptions about the nature of the constants $\alpha_{1,2}$ and $\beta_{1,2}$) use the conditions in Equation 5.4 to determine the values of the constants. We obtain

$$K_1 = \frac{2n+m}{2\left(n^2 + m^2 + nm\right)} b_1 + \frac{2m+n}{2\left(n^2 + m^2 + nm\right)} b_2 \qquad (5.5)$$

$$K_2 = \frac{mp}{2\left(n^2 + m^2 + nm\right)} b_1 - \frac{np}{2\left(n^2 + m^2 + nm\right)} b_2$$

In the direction of K_2, that is, the axial direction along the SWCNT, the effect of the quantization of the electron wavevector is only observed at extremely low temperatures [1] or in extremely short SWCNTs [2]. In the direction of K_1, on the other hand, the quantization of the electron wavevector is crucial. If we define λ as the de Broglie wavelength of the electron in the axial direction around the SWCNT, then we can describe the quantization as a requirement that the chiral vector must be an integer multiple of the wavelength: $C_h = \mu\lambda$ where μ is an integer. Recalling that C_h and K_1 are perpendicular allows us to simplify Equation 5.4 to $C_h K_1 = 2\pi$. We can then write an expression for the components of the electron wavevector k in the circumferential (k_R) and axial (k_T) directions:

$$k = k_R \hat{K}_1 + k_T \hat{K}_2$$

where

$$k_R = \frac{2\pi}{\lambda} = \frac{2\pi\mu}{C_h} = \mu K_1$$

Hence,

$$k = \mu K_1 + k_T \hat{K}_2 \qquad (5.6)$$

where
    k_T is a continuous variable
    $\hat{K}_2$ is a unit vector in the direction of K_2

The first Brillouin zone of the SWCNT is therefore a 1D line of finite length in reciprocal space, the values of k permitted by the conditions:

$$\mu = 0 \quad \text{and} \quad -\frac{K_2}{2} \le k_T \le \frac{K_2}{2}$$

Figure 5.2 shows the first Brillouin zone of the (5,5) and (10,10) SWCNTs, together with other permitted values of k,

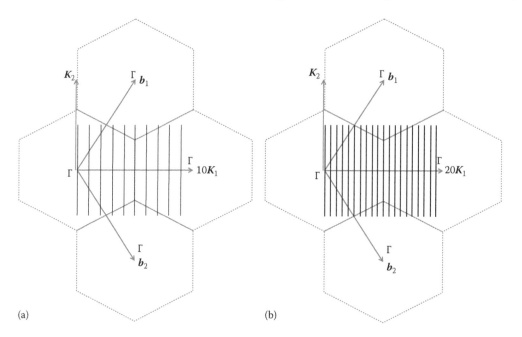

(a)   (b)

FIGURE 5.2
Permitted electron wavevectors (solid black lines) for the (5,5) (a) and (10,10) (b) SWCNTs, superimposed onto the graphene reciprocal lattice. The 1D lines of permitted wavevectors are separated by and are perpendicular to the SWCNT reciprocal lattice vector K_1. The first Brillouin zone is the line passing through the Γ point.

superimposed onto graphene's 2D reciprocal lattice. In both cases, we plot all permitted values of k necessary to reproduce the entire electronic dispersion relation of the SWCNT.

5.3 Prediction of the semiconducting or metallic nature of individual SWCNTs

We are now ready to predict whether an individual SWCNT is metallic or semiconducting using the 2D electronic dispersion relation of graphene (Figure 3.2). The SWCNT is metallic (i.e. a zero-bandgap semiconductor) if one of the lines of permitted values of k passes through the K point of the graphene electronic dispersion relation, where the conduction and valence bands meet. Whether or not this occurs depends on the angle at which the SWCNT axis cuts through the graphene lattice (in real space and in reciprocal space), and on the circumference of the SWCNT (which determines the scalar value of C_h and hence K_1). Of course, all of this information can be extracted from the (n,m) assignment; it is just a matter of doing the geometry.

Let us consider (Figure 5.3) the first Brillouin zone of graphene. The first Brillouin zone of the SWCNT passes through the Γ point and forms part of a line of permitted values of k that extend indefinitely in reciprocal space along the direction of K_2. We now draw a vector β perpendicular to this line of permitted k values linking it to the K point. When this vector is an integer multiple of K_1 one of the other lines of permitted values of k will pass through the K point. We also define the vector k_K leading from the Γ point to the K point

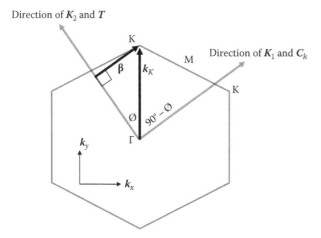

FIGURE 5.3
Conditions for an allowed value of the electron wavevector k to pass through the K point of the graphene reciprocal lattice: The vector β must be an integer multiple of K_1.

(for mathematical convenience we choose the K point directly above the Γ point along the y-axis):

$$k_K = \begin{pmatrix} 0 \\ \dfrac{4\pi}{3a} \end{pmatrix} \qquad (5.7)$$

To proceed, we begin by evaluating $C_h \cdot \hat{y}$ as follows:

$$C_h \cdot \hat{y} = C_h \cos\left(90° - \varnothing\right) = C_h \sin\varnothing = C_h \frac{\beta}{k_K}$$

We can substitute for k_K using Equation 5.7 and can evaluate $C_h \cdot \hat{y}$ using Equations 1.2 and 1.5, which leads us to

$$\frac{a}{2}\left(n - m\right) = C_h\beta \times \frac{3a}{4\pi}$$

Since $C_h = \dfrac{2\pi}{K_1}$ we obtain

$$\beta = \frac{\left(n - m\right)K_1}{3} \qquad (5.8)$$

Hence, for the SWCNT to be metallic, we require that $(n - m)/3$ is an integer to ensure that β is an integer multiple of K_1. Hence, all armchair ($n = m$) SWCNTs are metallic, and zigzag ($m = 0$) SWCNTs are metallic whenever n is divisible by 3. We can plot in diagrammatic form which SWCNTs are metallic by plotting the values of C_h in the graphene lattice (Figure 5.4).

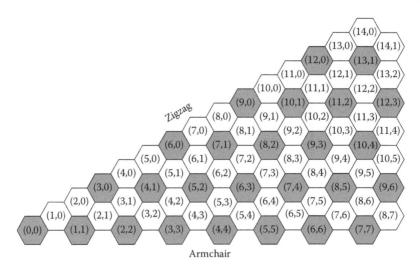

FIGURE 5.4
Illustration of semiconducting or metallic nature of different SWCNTs. (n,m) indices indicate (on the graphene lattice) the chiral vector of each SWCNT, and shaded hexagons indicate metallic SWCNTs, whilst open hexagons indicate semiconducting SWCNTs, as indicated by Equation 5.8.

5.4 Energy dispersion relation of armchair SWCNTs

We will now derive specific energy dispersion relations for armchair and zigzag SWCNTs. For the case of armchair SWCNTs ($n = m$), we can write (using Equations 5.5 and 1.2)

$$K_1 = \frac{1}{2n}(b_1 + b_2) = \begin{pmatrix} \dfrac{2\pi}{\sqrt{3}an} \\ 0 \end{pmatrix} \qquad (5.9)$$

$$K_2 = \frac{1}{2}(b_1 - b_2) = \begin{pmatrix} 0 \\ \dfrac{2\pi}{a} \end{pmatrix}$$

Utilizing Equation 5.6 we can then write an expression for k, the permitted values of the electron wavevector:

$$k = \begin{pmatrix} \dfrac{2\mu\pi}{\sqrt{3}an} \\ k_T \end{pmatrix}$$

and hence an expression for $w(k)$ (substituting into Equation 3.11)

$$w(k) = \sqrt{1 + 4\cos\left(\frac{\mu\pi}{n}\right)\cos\left(\frac{k_T a}{2}\right) + 4\left(\cos\left(\frac{k_T a}{2}\right)\right)^2}$$

Using our expression for the energy dispersion relation (Equation 3.10, with $\varepsilon_{2p} = 0$), we can then write the electronic dispersion relation for an armchair SWCNT. In the first Brillouin zone, $|k_T| \leq (K_2/2)$ allows us to define the dispersion relation as follows:

$$-\frac{\pi}{a} \leq k_T \leq \frac{\pi}{a} \qquad (5.10)$$

$$E(k) = \pm\frac{tw(k)}{1 \pm sw(k)}$$

Note that each permitted value of μ leads to a different branch of the dispersion relation. To obtain the entire dispersion relation we need to plot $E(k)$ for all values of μ up to μ_T, the value for which the vector $\mu_T K_1$ is an integer multiple of the graphene reciprocal lattice vectors. Referring to Equation 5.9, we can observe that this condition is met when $\mu_T = 2n$.

In Figure 5.2, we plotted out all the permitted values of electron wavevector necessary to obtain the entire electronic dispersion relation for the (5,5) and (10,10) armchair SWCNTs. We can now, using Equation 5.10, plot the dispersion relation.

The dispersion relation for the (5,5) SWCNT has 10 branches and that for the (10,10) SWCNT has 20 branches. Note that the first Brillouin zone as we have defined it here ($\mu = 0$ and hence $k = k_T \hat{K}_2$) only leads to one branch of the dispersion relation; that is, it does not fulfil the conventional definition of the first Brillouin zone. For this reason, some texts choose instead to define the first Brillouin zone as all values of k permitted for $\mu = 0$ to $\mu = \mu_T$. However, this approach also has problems, as it is not consistent with the definition of K_1 as the reciprocal lattice vector.

Figure 5.5 shows the complete energy dispersion relation for (5,5) and (10,10) armchair SWCNTs. There is a twofold degeneracy of all branches of the dispersion relation except that which passes through the Γ point (leading to the highest and lowest energy branches) and that which passes through the K and M points.

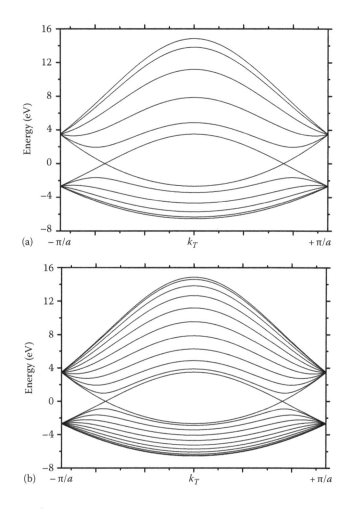

FIGURE 5.5
Energy dispersion relation for armchair (5,5) (a) and (10,10) (b) SWCNTs.

5.5   Energy dispersion relation of zigzag SWCNTs

Analogous to the case of armchair SWCNTs, we can evaluate Equation 5.5 for K_1 and K_2 for the case of the zigzag $(n,0)$ SWCNT:

$$K_1 = \frac{1}{n}b_1 + \frac{1}{2n}b_2 = \begin{pmatrix} \dfrac{\sqrt{3}\pi}{na} \\ \dfrac{\pi}{na} \end{pmatrix} \tag{5.11}$$

$$K_2 = -\frac{1}{2}b_2 = \begin{pmatrix} -\dfrac{\pi}{\sqrt{3}a} \\ \dfrac{\pi}{a} \end{pmatrix}$$

Figure 5.6 plots the reciprocal lattice vectors for the (5,5) armchair SWCNT (Equation 5.9) and the (10,0) zigzag SWCNT (Equation 5.11). In the case of the armchair SWCNT, the reciprocal lattice vectors K_1 and K_2 pointed along the x- and y-axis, respectively. In the expressions for the zigzag SWCNT, this is no longer the case, so inserting the expressions in Equation 5.11 directly into Equations 3.10 and 3.11 would yield a somewhat lengthy expression for the energy.

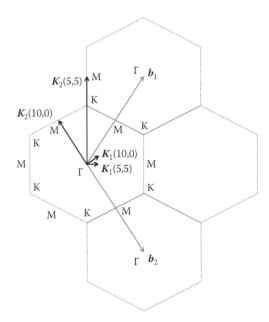

FIGURE 5.6
The reciprocal lattice vectors K_1 and K_2 for an armchair (5,5) SWCNT and zigzag (10,0) SWCNT plotted on the graphene reciprocal lattice.

We therefore observe that if we rotate K_1 and K_2 in the zig-zag (10,0) SWCNT by 60° anticlockwise, due to the symmetry of the graphene lattice we trace exactly the same path through reciprocal space: From the Γ point towards the K point in the case of K_1 and from the Γ point towards the M point in the case of K_2. If we perform this operation we will hence obtain an energy dispersion relation which is identical as a function of μ and k_T. We therefore rotate K_1 and K_2 to obtain the following expressions:

$$K_1' = \begin{pmatrix} 0 \\ \dfrac{2\pi}{na} \end{pmatrix} \tag{5.12}$$

$$K_2' = \begin{pmatrix} -\dfrac{2\pi}{\sqrt{3}a} \\ 0 \end{pmatrix}$$

We hence obtain

$$k = \mu K_1' + k_T \hat{K}_2' = \begin{pmatrix} k_T \\ 2\pi\mu \\ \overline{na} \end{pmatrix} \tag{5.13}$$

and

$$E(k) = \pm \frac{tw(k)}{1 \pm sw(k)} \tag{5.14}$$

where

$$w(k) = \sqrt{1 + 4\cos\left(\frac{\sqrt{3}k_T a}{2}\right)\cos\left(\frac{\pi\mu}{n}\right) + 4\left(\cos\left(\frac{\pi\mu}{n}\right)\right)^2}$$

$$-\frac{\pi}{\sqrt{3}a} \leq k_T \leq \frac{\pi}{\sqrt{3}a}$$

Referring to Equation 5.11 we can see again that μK_1 and hence $\mu K_1'$ is an integer combination of b_1 and b_2 when μ = 2n, so there are again 2n branches to the dispersion relation, which is plotted for the (6,0) and (10,0) zigzag SWCNTs in Figure 5.7. However, there is a twofold degeneracy of all bands except those passing through the Γ point. There are hence 7 separate energy bands in the case of the (6,0) SWCNT and 11 separate energy bands in the case of the (10,0) SWCNT.

Note that, in the case of the semiconducting (10,0) SWCNT in Figure 5.7, and other semiconducting SWCNTs, there exists a direct extremal electronic bandgap with a flat dispersion relation at this point. We thus expect a large density of states above and

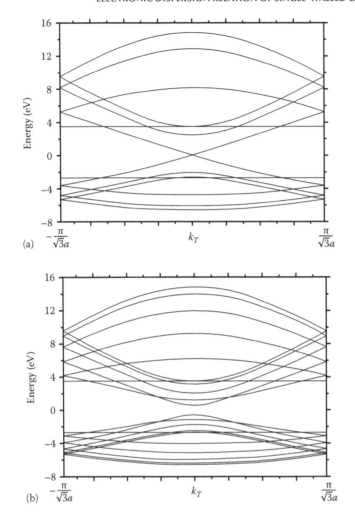

FIGURE 5.7
Energy dispersion relation for zigzag (6,0) (a) and (10,0) (b) SWCNTs. The (6,0) dispersion relation includes a line of allowed wavevectors passing through the K point, ensuring that the tube is metallic. The (10,0) SWCNT on the other hand has a small direct extremal bandgap.

below the extremal bandgap, the observation of photoluminescence (PL) and potential applications in optoelectronics. These properties are a remarkable contrast to those of graphene, and we thus discuss them in some detail in the following section.

5.6   Electronic density of states in SWCNTs

In graphene, the electronic density of states drops to virtually zero at the K point where the conduction and valence bands meet. In SWCNTs however, all branches of the dispersion relation except the branch in metallic SWCNTs that intersects the K point exhibit maxima and minima at which the density of states is extremely large. These spikes in the density of states are known as van Hove singularities. They are responsible for the observation of strong PL from isolated semiconducting SWCNTs and the potential applications of SWCNTs in optoelectronics.

5.6.1   Density of states in one-dimensional (1D) systems

To characterize the van Hove singularities and optical properties of SWCNTs we need to begin by obtaining an expression for the density of quantum states available to a 1D system. This is analogous to our treatment of density of states in a 2D system in Section 3.7.1.

The solutions to the Schrödinger equation for a particle confined in a 1D infinite square well ([7] of Chapter 2) are wave functions of form $\psi(k) = A \sin k_x x$. We require that $\psi(k) = 0$ at the boundaries of the square well. Hence $k_x L = n_x \pi$, where n_x is an integer.

The allowed values of k form a line in k-space and the spacing between adjacent lattice points is (π/L). Hence, the number of lattice points per unit length in k-space is (L/π), and the number N_{kr} of lattice points per unit length in k-space and per unit length in real space is

$$N_{kr} = \frac{1}{\pi} \tag{5.15}$$

Let us now consider the available quantum states $n(k_0)$ for an electron with wavevector up to a certain value k_0:

$$n(k_0) = 2 \times \frac{1}{2} \times 2k_0 \times \frac{1}{\pi} \tag{5.16}$$

Here, we have the factor of 2 to account for the two possible orientations of the electron spin (just as we did for the 2D case) and a factor to avoid double-counting the number of available states, which of 1/2 this time. Then we have the length of the line in k-space, $2k_0$, and the number of quantum states available per unit length $(1/\pi)$.

The density of states $\rho(k_0)dk$, the number of quantum states available between wavevectors k_0 and $(k_0 + dk)$, per unit length (of real space) is therefore

$$\rho(k_0)dk = \frac{dn}{dk_0} dk = \frac{2}{\pi} dk \tag{5.17}$$

We then obtain the density of states as a function of energy by writing $\rho(E)|dE| = \rho(k_0)|dk|$ which gives

$$\rho(E)dE = \rho(k_0)\left|\frac{dE}{dk}\right|^{-1} dE = \frac{2}{\pi}\left|\frac{dE}{dk}\right|^{-1} dE \tag{5.18}$$

5.6.2   Density of states in SWCNTs

We can use Equation 5.18 to calculate the electronic density of states as a function of energy $\rho(E)$ using the dispersion relation $E(k)$. Each of the permitted quantized values of the electron wavevector in the circumferential direction (i.e. each integer value of μ in Equation 5.6 onwards) results in a different branch of the dispersion relation. We may therefore label these branches as $E(\mu, k_T)$, where k_T is the component of the electron wavevector in the axial direction, which can vary continuously. Hence,

$$k(\mu, k_T) = \mu K_1 + k_T \hat{K}_2 \qquad (5.19)$$

$$E(\mu, k_T) = \pm \frac{tw\big(k(\mu, k_T)\big)}{1 \pm sw\big(k(\mu, k_T)\big)}$$

We can obtain the total density of states as a function of energy by summing the expression in Equation 5.18 over all branches of the dispersion relation (i.e. all values of μ):

$$\rho(E)dE = \frac{2}{\pi} \sum_{\mu} \left| \frac{dE(\mu, k_T)}{dk} \right|^{-1} dE \qquad (5.20)$$

In Equation 5.20, we must perform the summation over all values of μ that produce an energy dispersion relation in the energy range in which we are interested. By inspection of Equation 5.20, we can see that when $(dE/dk) \to 0$ the density of states $\rho(E)$ becomes extremely large. Every maximum and every minimum in each branch of the dispersion relation leads to a large positive spike in the density of states. These spikes are known as van Hove singularities.

Figure 5.8 shows the density of states calculated using Equation 5.20 for the semiconducting (10,0) SWCNT. As we can see, there is a large density of states on either side of an extremal bandgap of about 1.1 eV. This condition allows for strong PL at around 1100 nm and potential applications for SWCNTs in optoelectronics. Each E_{ii} label corresponds to an energy difference between van Hove singularities in the valence band and conduction band, sharing the same value of μ. As we shall see later, these are the transitions we can excite optically.

It is possible to construct a plot of these different allowed optical transition energies in SWCNT as a function of the SWCNT diameter for all tubes with diameters within a certain range; such a plot is referred to as a Kataura plot ([8] of Chapter 1 and [3]). We will show in Chapter 7 how such a plot can be used to assign (n,m) indices to an SWCNT using Raman spectroscopy. Figure 5.9 shows an example Kataura plot, the theoretically predicted E_{ii} values for SWCNT up to 3.0 nm diameter ([8] of Chapter 1).

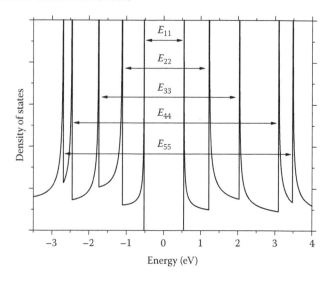

FIGURE 5.8
Electronic density of states for the semiconducting (10,0) zigzag SWCNT as a function of energy. Each spike (van Hove singularity) originates from a maximum or minimum in a different branch of the dispersion relation (Figure 5.7). Each label E_{ii} corresponds to the energy of an optically allowed transition.

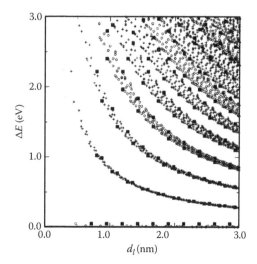

FIGURE 5.9
Theoretically predicted optical resonance energies $\Delta E = E_{ii}$ for SWCNTs up to 3.0 nm diameter. The data points at $\Delta E = 0$ eV correspond to metallic SWCNTs. The lowest curved line of points corresponds to E_{11} transitions, the next lowest to E_{22} transitions, etc. (Reprinted with permission from Saito, R., Dresselhaus, G. and Dresselhaus, M.S., *Phys. Rev. B*, 61, 2981. Copyright 2001 by the American Physical Society.)

5.7   Excitons in SWCNTs

In our treatment so far of the electronic dispersion relations of both graphene and SWCNTs, we have neglected any possibility of interactions between electrons in the conduction band(s) and holes in the valence band(s). We have assumed that if we excite an electron into the conduction band, leaving behind a hole in the valence band, the only possible interaction between the electron and hole is a complete recombination in which the electron rejoins the valence band. Unless this occurs, the energy of the electron is given according to the dispersion relation for the conduction band regardless of the presence (or otherwise) of the hole.

This assumption is justified for graphene but not for SWCNTs. Electrons and holes can interact without recombining (in SWCNTs and in many other important semiconductors [4]), forming a system called an exciton. We can understand the basics of exciton behaviour by considering the electron–hole system as a hydrogenic atom with energy levels given by the familiar Bohr formula, except that the mass is the effective mass of the electron–hole system. Thus the exciton energy levels depend strongly on the nature of the dispersion relations for electrons and holes as this is what determines the effective mass.

We may apply the Bohr formula to describe the exciton binding energy E_n as follows [4]:

$$E_n = \frac{\mu e^4}{8h^2 \varepsilon_r^2 \varepsilon_0^2 n^2} \tag{5.21}$$

where
    the integer n labels the different energy levels
    ε_r is the relative permittivity in the environment of the exciton
    μ is the reduced effective mass of the exciton system, obtained from the effective masses of the electron (m_e^*) and hole (m_h^*)

$$\mu = \frac{m_e^* m_h^*}{m_e^* + m_h^*} \tag{5.22}$$

The effective masses of the electrons and holes are derived from the dispersion relation $E(k)$ ([2] of Chapter 1):

$$m_{e/h}^* = \hbar^2 \left[\frac{d^2 E}{dk^2} \right]^{-1} \tag{5.23}$$

with an equivalent relation applying for the holes. As we shall see in the next section, the presence of excitons is necessary to explain the relation between excitation and emission photon energy in the PL experiments conducted on SWCNT.

5.8   Experimental verification of SWCNT dispersion relation

The SWCNT dispersion relation predicted using the zone-folding procedure mentioned earlier is based on the graphene dispersion relation. Since 2004, the electronic dispersion relation of graphene itself has been verified using a number of different experimental methods (outlined in Chapters 3, 4 and 7). However, a diverse range of measurements have also been made directly on the electronic dispersion relation of SWCNT, verifying the efficacy of the zone-folding procedure and the importance of exciton formation in determining the exact bandgap energy.

Scanning tunnelling microscopy and spectroscopy experiments have confirmed the predicted existence of metallic or semiconducting behaviour in SWCNTs as a function of the chiral indices [5] and also demonstrated the existence of van Hove singularities in the electronic density of states via the electric field effect.

The optical properties of SWCNT also depend sensitively on the electronic dispersion relation. We can excite an electron from the valence band to the conduction band through photon absorption (in a PL experiment) or inelastic scattering of a photon (in a Raman spectroscopy experiment). The photon has negligible momentum compared to an electron at any point other than the Γ point (Appendix A). Hence, in both cases we cannot cause any change in the electron wavevector during the excitation process. From a given van Hove singularity in the valence band, we can only excite to the van Hove singularity in the conduction band which has the same value of μ. These are the optically allowed transitions labelled E_{ii} in Figure 5.8.

Due to the resonant nature of Raman scattering in SWCNTs, it is possible to observe a Raman spectrum from a single SWCNT when the excitation laser energy coincides with the energy gap between two of the van Hove singularities. Combining this information with the measurement of tube diameter provided by the radial breathing Raman mode, it is possible to use Raman spectroscopy to identify the (n,m) indices of an SWCNT [6].

PL has also been observed from SWCNT, providing direct and unambiguous evidence for the large density of states on either side of the extremal bandgap in semiconducting SWCNTs and the importance of excitons in the system. We shall review Raman spectroscopy of SWCNTs in Chapter 7, scanning tunnelling microscopy and spectroscopy of SWCNTs in Chapter 8 and PL measurements on SWCNTs in the following section of this chapter.

5.8.1   Photoluminescence (PL) in SWCNTs

Despite the presence of extremal electronic bandgaps in the visible and near-IR range, PL experiments on SWCNT initially proved challenging to perform, due to non-radiative recombination. Whilst emission of light from SWCNT deposited on a substrate is observed [7,8], non-radiative recombination has prevented the observation of strong and sharp PL peaks from SWCNT contained in bundles (due to the presence of metallic SWCNTs) or deposited on a substrate [9,10].

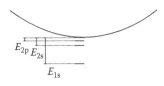

The observation of strong PL across the direct electronic bandgaps in SWCNTs was made possible through the isolation of individual SWCNTs in solution by breaking apart the SWCNT bundle using ultrasound [9] as described in Chapter 9. The principal scientific advance made possible by the observation of PL was the understanding that the emission of light from SWCNT originates from the recombination of electrons and holes that are already strongly correlated due to being bound in an exciton [11].

FIGURE 5.10

Sketch of low-lying excitonic energy levels near the extremal bandgap in a semiconducting SWCNT.

To understand this, we can visualize the various excitonic energy levels lying just below the lowest point of the conduction band (Figure 5.10). Just as in the real hydrogenic atom, the different energy levels involve the system having different orbital angular momentum and therefore different parity. Since parity is conserved in the electromagnetic interaction, the parity of the different levels determines which of the levels can be populated when an electron from the valence band absorbs one photon and which can be populated when an electron from the valence band absorbs two photons simultaneously. As far as the lowest levels are concerned, it is possible to populate the 1s level with single photon absorption and the 2p level with two photon absorption. A detailed discussion of why these specific transitions are those permitted by the conservation of parity would be beyond the scope of this text, but the interested reader is referred to more detailed works on this topic [11,12].

If the observed PL from SWCNTs was from electron–hole recombination directly across the bandgap E_{ii} (i.e. exciton formation played no role) then we would expect to observe PL at energy E_{ii} as when the total energy provided by the incident photon(s) reaches E_{ii} – regardless of whether this energy comes from one photon or two photons.

However, what we actually observe is PL following one-photon excitation at significantly lower energy, $E_{ii} - E_{1s}$, when the incident photon energy reaches $E_{ii} - E_{1s}$. Furthermore, in the case of two-photon luminescence (Figure 5.11), the PL is not observed until the total energy absorbed from both incident photons is $E_{ii} - E_{2p}$, significantly higher than $E_{ii} - E_{1s}$. The PL is then observed at energy $E_{ii} - E_{1s}$ as it is preceded by non-radiative decay of the exciton into the 1s energy level. These observations demonstrate the importance of excitons in determining the energies at which light is emitted from SWCNTs. Exciton binding energies are significant: up to 0.4 eV [11]. It is also known that the exciton binding energy is dependent on the environment of the SWCNT, or more specifically the

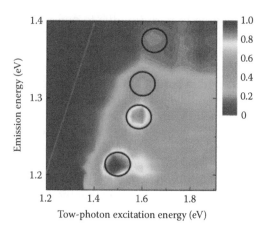

FIGURE 5.11

PL intensity from single-walled carbon nanotube (SWCNT) sample as a function of two-photon excitation energy. Peaks correspond to emission from SWCNTs within the sample with different (n,m). Black line indicates emission energy being equal to the two-photon excitation energy. Note that in all cases the emission is at lower energy. (From Wang, F. et al., *Science*, 308, 838, 2005. Reprinted with permission of AAAS.)

permittivity of the environment (referring to Equation 5.21) [10,11,13–15]. This is because the electric field lines between the electron and hole forming the exciton extend well outside of the SWCNT.

Figure 5.11 demonstrates the observed dependence of PL intensity observed from an SWCNT sample on the excitation energy for the case of two-photon excitation. The various emission peaks correspond to the total two-photon excitation energy coinciding with the energy required to excite PL in different SWCNT within the sample. Note that in all cases the PL observed is at significantly lower energy than the excitation energy. This is the key observation which can only be explained by the formation of excitons.

In addition to PL, electroluminescence [8] and photoconductivity [16] have been observed in SWCNT. The optical transition energies have been thoroughly catalogued [17], and their measurement (using PL and other optical techniques) has been used as a method to identify the (n,m) indices in SWCNT (see, for instance, Reference 18). However, the exciton field lines extend outside of the SWCNT and hence the permittivity of the SWCNT environment can cause the transition energies to shift [10,11,13–15]. It is therefore prudent to back up (n,m) assignments made using measurements of the optical transition energies with some other method, and/or to ensure that the reference data used to make the (n,m) assignment come from SWCNTs in the same environment as the sample.

5.9   Curvature effects in SWCNTs

The zone-folding method that we have used so far to predict the electronic dispersion relation for SWCNT assumes that the only change induced in the graphene dispersion relation by rolling up the graphene sheet into a tube is the quantum confinement of electrons in the direction around the tube axis, and restriction on allowed wavevectors for the electrons that results.

The experimental evidence indicates that this assumption is generally reasonable. With this assumption we can accurately predict which SWCNT are semiconducting and which are metallic, and also predict the existence of the observed van Hove singularities in the electronic density of states.

However, the zone-folding approximation is not perfect because it uses the results (Equations 5.10 and 5.14) of the tight-binding calculation of the graphene electronic dispersion relation that we performed in Chapter 3. This calculation assumes that all carbon atoms lie in the same plane (Equation 3.7) and assumes perfect sp^2 hybridization as a result. We assume that the π orbitals take no part in hybridization. This is no longer the case if the graphene lattice is rolled up into a tube, because the π orbitals must take part in hybridization to ensure the bonds point in the right direction to allow the curvature. Instead of sp^2 hybridization we have $sp^{2+\eta}$ hybridization (Section 2.4).

In addition, the length of bonds in the direction around the circumference of the SWCNT gets slightly smaller as a result of rolling up the graphene sheet into a tube. As a result, the lattice no longer possesses the perfect hexagonal symmetry of the graphene lattice.

A detailed consideration of these effects would be beyond the scope of this introductory text, so the reader is referred to more detailed works on the subject ([17] of Chapter 2 and [19,20]). The clearest effect of curvature, however, is the opening of a small (<0.1 eV) electronic bandgap in those zigzag SWCNTs which we would otherwise expect to be metallic (i.e. zero-bandgap semiconductors) [20].

References

1. SJ Tans et al., *Nature* **386**, 474 (1997).

2. HWCh. Postma, T Teepen, Z Yao, M Grifoni and C Dekker, *Science* **293**, 76 (2001).

3. H Kataura et al., *Synth. Met.* **103**, 2555 (1999).

4. PY Yu and M Cardona, *Fundamentals of Semiconductors*, Springer, Berlin (1996).

5. JWG Wildöer, LC Venema, AG Rinzler, RE Smalley and C Dekker, *Nature* **391**, 59 (1998).

6. A Jorio et al., *Phys. Rev. Lett.* **86**, 1118 (2001).

7. P Avouris, *MRS Bull.* **29**, 403 (June 2004).

8. JA Misewich, R Martel, P Avouris, JC Tsang, S Heinze and J Tersoff, *Science* **300**, 783 (2003).

9. MJ O'Connell et al., *Science* **297**, 593 (2002).

10. P Avouris, M Freitag and V Perebeinos, *Nat. Photonics* **2**, 341 (2008).

11. F Wang, G Dukovic, LE Brus and TF Heinz, *Science* **308**, 838 (2005).

12. A Shimizu, T Ogawa and H Sakaki, *Phys. Rev. B* **45**, 11338 (1992).

13. RS Deacon, K-C Chuang, J Doig, B Mortimer and RJ Nicholas, *Phys. Rev. B* **74**, 201402(R) (2006).

14. M Steiner et al., *Appl. Phys. A* **96**, 271 (2009).

15. PT Araujo, PBC Pesce, MS Dresselhaus, K Sato, R Saito and A Jorio, *Physica E* **42**, 1251 (2010).

16. M Freitag, Y Martin, JA Misewich, R Martel and P Avouris, *Nano Lett.* **3**, 1067 (2003).

17. K Liu et al., *Nat. Nanotechnol.* **7**, 325 (2012).

18. K Liu et al., *Nat. Nanotechnol.* **8**, 917 (2013).

19. S Reich, C Thomsen and J Maultzsch, *Carbon Nanotubes: Basic Concepts and Physical Properties*, Wiley-VCH, Chichester (2004).

20. CL Kane and EJ Mele, *Phys. Rev. Lett.* **78**, 1932 (1997).

6 Phonons in Graphene and Single-Walled Carbon Nanotubes (SWCNTs)

6.1 Why study the phonons?

The behaviour of both acoustic and optical phonons in graphene and single-walled carbon nanotubes (SWCNTs) forms an important part of our understanding of these materials and ability to characterize them. For instance, it is the behaviour of the low-frequency, out-of-plane acoustic phonons in graphene that leads to the material having a negative thermal expansion coefficient (Chapter 10), and the observation of several Raman-active phonons in graphene and SWCNTs that allows the use of Raman spectroscopy as a quick and non-destructive method to measure parameters such as the thickness of graphene samples and the diameter of SWCNTs (Chapter 7).

6.2 Theory

For decades prior to the isolation of graphene, the phonon modes in graphene have been studied theoretically as these are also the in-plane phonon modes in graphite. The phonon dispersion relation has been predicted using traditional semi-classical "ball and spring" force constant models (for instance, References 12 of Chapter 1 and [1,2]). More recently, it has also been studied using density functional theory (DFT) calculations, made possible by modern computing facilities (for instance, References 3–5). The results of the DFT calculations are in reasonable agreement with the results obtained in previous decades using the ball and spring models.

Figure 6.1 shows the phonon dispersion relation of graphene as predicted by DFT calculations [3] whilst travelling between the various high symmetry points in the graphene first Brillouin zone. Locations where the phonon dispersion curve passes through the high symmetry points are labelled, and in Figure 6.2 the phonon eigenmodes at these high symmetry points [5] are shown.

In common with phonons in other crystalline solids, we can divide the phonons in graphene into two categories: acoustic phonons and optical phonons. Acoustic phonons are so called because they are responsible for the propagation of sound waves through solids. Acoustic phonons are due to atoms vibrating in phase, whilst optical phonons are due to

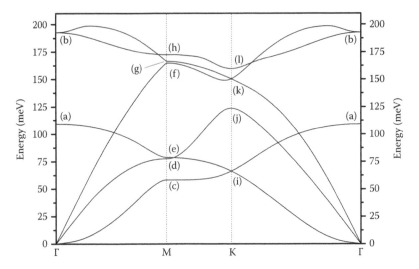

FIGURE 6.1
Ab initio prediction of the phonon dispersion relation for graphene, labelled at various high symmetry points. (Reprinted with permission from Mounet, N. and Marzari, M., *Phys. Rev. B*, 71, 205214. Copyright 2005 by the American Physical Society.)

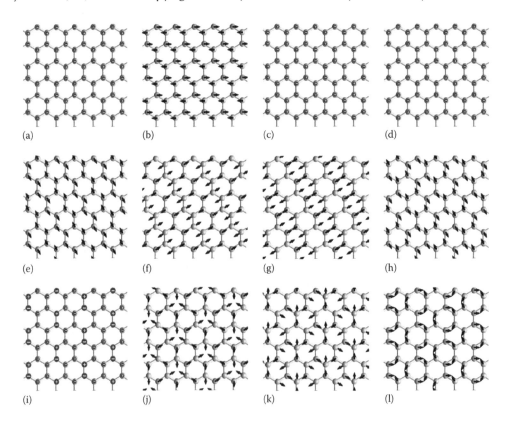

FIGURE 6.2
Eigenmodes of different phonons at the high symmetry points (see Figure 6.1) in the graphene first Brillouin zone [5]. Arrows indicate movement of atoms within the graphene plane, and ± signs indicate movement in the out-of-plane direction.

atoms vibrating out of phase. Optical phonons are so called because the out-of-phase vibration leads to the polarizability of the crystal changing and the possibility to excite these phonons using electromagnetic radiation.

The out-of-phase vibration necessitates stretching and compressing the interatomic bond to a far greater degree than the in phase vibrations and therefore, in graphene as in other crystalline solids, the optical phonons have higher energy. Referring to Figure 6.1, the three different phonon modes originating from the Γ point at zero energy are acoustic phonons, but the eigenmodes labelled (a)-(l) and shown in Figure 6.2 are all optical phonons. We focus on the acoustic phonons in this chapter, and focus on the optical phonons in Chapter 7 where we see how certain of these phonons can be excited using Raman spectroscopy. Readers are also referred to Appendix A for further discussion on the distinction between optical and acoustic phonons, and which phonons can be excited using Raman spectroscopy.

6.2.1   Low-energy acoustic phonons near the Γ point

We pay particular attention to the three acoustic phonon dispersion curves emanating from $E = 0$ at the Γ point. Figure 6.3 shows the low-energy region of the phonon dispersion relation in the area between the Γ point and M point. Out of the three phonon dispersion curves originating from the Γ point, the lowest-energy mode is a mode in which the atoms oscillate in the out-of-plane direction whilst the other two modes are both in-plane modes ([12] of Chapter 1 and [3,5]). We will concentrate particularly on the lowest-energy, out-of-plane mode.

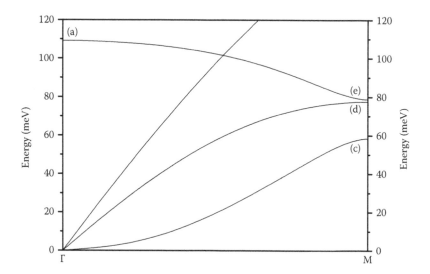

FIGURE 6.3

Enlargement of phonon dispersion relations of graphene [3] at low energy between the Γ point and the M point.

At the Γ point, the mode has zero energy as well as zero wavevector since all atoms are moving out-of-plane in the same direction so there is no actual vibration. At the M point (c) this mode results in ripples forming in the graphene sheet with wavelength (in real space) of double the lattice constant a. As we move from the Γ point to the M point the wavelength of these ripples gradually decreases from infinity to $2a$.

In the vicinity of the Γ point, the two in-plane modes both exhibit a linear relationship between frequency (energy) and wavevector ($\omega \propto k$), whilst the out-of-plane mode exhibits a quadratic relationship ($\omega \propto k^2$). The linear dispersion relation for the in-plane modes in the vicinity of the Γ point is what is normally observed for longitudinal and transverse phonons. It is also what we expect to observe for vibrations in a hypothetical 1D monatomic or diatomic chain and for the propagation of longitudinal waves in a continuous homogeneous solid ([2] of Chapter 1).

The quadratic dispersion relation for the out-of-plane acoustic phonon, however, is unusual (but not unique [6,7]). In this case, it is a result of the specific symmetry of the graphene lattice ([12] of Chapter 1) and does not occur, for instance, in diamond [3].

The existence and nature of the out-of-plane acoustic phonon has two important consequences which are explored in Chapter 10.

First, the fact that it has significantly lower energy than the in-plane phonons ensures that at low and ambient temperature, there are significantly more quantum states available for the out-of-plane phonon than the in-plane phonons. This results in graphene exhibiting a negative thermal expansion coefficient under these conditions.

Second, we need to know the nature of the dispersion relation for these phonons ($\omega \propto k^2$, $E \propto k^2$) in order to evaluate the number of phonons thermally excited at a given temperature. We do this in Chapter 10 in order to understand the thermodynamic stability (or lack thereof) of graphene.

6.3   Experiment

In Figure 6.1, we showed the results of a theoretical prediction of the phonon dispersion relation in graphene. Unfortunately, it is only possible to directly measure a very limited part of this dispersion relation using Raman spectroscopy as discussed in Chapter 7 and (much less commonly) using infrared spectroscopy [8].

However, we expect the dispersion relation for phonons in graphene to be very similar to that for phonons propagating in the in-plane direction in graphite [3]. Due to the larger sample size, the in-plane phonons in graphite have been studied using several experimental techniques: Inelastic X-ray scattering [5], electron energy loss spectroscopy [9] and inelastic neutron scattering [2] have been utilized to map out the entire phonon dispersion relation. The results agree closely with what is expected theoretically for both graphite and graphene.

6.4 Phonons in SWCNTs

6.4.1 General comments

The phonon dispersion relation in an SWCNT is different and far more complex than the phonon dispersion relation in graphene. It would be beyond the scope of this introductory text to provide a full treatment of the phonon dispersion relation in SWCNTs. For this, the reader is referred to various more detailed texts on the subject ([12] of Chapter 1, [19] of Chapter 5 and [10]).

Here, we will, however, discuss key changes in phonon behaviour caused by the rolling up of the graphene sheet into a tube in order to inform our discussion in Chapter 7 of the Raman spectra of SWCNTs.

The first point to note is that the phonon dispersion relation for SWCNT has many more branches than the phonon dispersion relation for graphene. In graphene (Figure 6.1), there are very few phonon eigenmodes with non-zero energy at the Γ point; whilst in a typical SWCNT, we expect upwards of 25 phonon eigenmodes to be present at the Γ point. This is because, when we roll the graphene sheet up into an SWCNT, we increase the size of the (real space) primitive unit cell by at least an order of magnitude. In graphene the unit cell consists of two atoms, whilst in an SWCNT the primitive unit cell extends all the way around the nanotube and a considerable distance along the nanotube axis (defined by Equations 5.1 and 5.2). The effect of this increase in the unit cell size is to increase the number of branches to the phonon dispersion relation; in the previous chapter, we observed how the increase in unit cell size has an analogous effect on the electronic dispersion relation.

The "zone-folding" procedure that we used in the previous chapter to derive the electronic dispersion relation in SWCNTs from the graphene electronic dispersion relation can also be utilized to derive the phonon dispersion relation for SWCNTs from the phonon dispersion relation for graphene shown in Figure 6.1. Just as when applied to the electronic dispersion relation, the zone-folding procedure considers the quantization of phonon wavevector in the circumferential direction around the nanotube. As a result, the 2D dispersion relation for graphene changes to a 1D dispersion relation giving the phonon energy as a function of the value of the phonon wavevector along the nanotube axis (shown in Figure 6.4 for the (4,4) SWCNT). A large number of branches to this dispersion

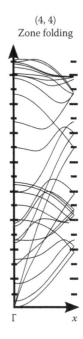

(4, 4)
Zone folding

Γ   x

FIGURE 6.4
1D phonon dispersion relation for (4, 4) single-walled carbon nanotube (SWCNT) obtained using the zone-folding method. The x-axis denotes the component of the phonon wavevector parallel to the SWCNT axis, whilst each branch corresponds to a different quantized value of the wavevector in the circumferential direction. (Reprinted with permission from Sánchez-Portal, D., Artacho, E., Soler, J.M., Rubio, A. and Ordejón, P., *Phys. Rev. B*, 59, 12678. Copyright 2009 by the American Physical Society.)

relation are due to the different quantized values that the component of the phonon wavevector in the circumferential direction around the nanotube can take.

Just as with the phonon dispersion relation of graphene presented in Figure 6.1, it is only possible to directly measure the frequencies of a small proportion of the phonons expected to be present in the SWCNT, at certain discrete points in the Brillouin zone. This is done using Raman spectroscopy (Chapter 7) or (less commonly) using infrared spectroscopy (see discussion in Reference 19 of Chapter 5). Unlike graphene, however, the phonon dispersion relation of SWCNT is not expected to be similar to the extremely well-characterized material graphite, so there is thus more uncertainly regarding the exact nature of the phonon dispersion relation away from the Γ point.

In addition, there are two additional low-energy phonon modes expected in SWCNT, the existence of which is not predicted by the zone-folding approach outlined earlier [11]. One of these modes (the radial breathing mode [RBM]) can be measured using Raman spectroscopy (is Raman-active) and (as we shall see in the next chapter) plays a major role in characterization of SWCNT. It is therefore discussed in the following section.

6.4.2 Low-energy phonon: Radial breathing mode

The low-energy out-of-plane phonon mode in graphene discussed in Section 6.2.1 behaves very differently in an SWCNT. In graphene, as shown in Figure 6.3, the mode has zero energy at the Γ point as it corresponds to all atoms moving in the same direction (along the z-axis, out of the graphene plane). No bonds expand, contract or bend at all under such a movement so there is no vibrational mode.

But in an SWCNT, a movement of all atoms away from the tube axis along the direction out of the graphene plane results in an expansion of the entire tube and stretching of the C–C bonds. This mode does therefore result in a vibration at the Γ point with non-zero energy and frequency. This is a change which is not predicted by the zone-folding approach to the SWCNT phonon dispersion relation outlined in the previous section, but which is observed experimentally: This mode (as we shall see in the next chapter) can be measured with Raman spectroscopy and is referred to as the radial breathing mode (RBM). Figure 6.5 illustrates the zero-frequency eigenmode at the Γ point in a flat graphene sheet and the non-zero frequency eigenmode which this corresponds to in an SWCNT.

The frequency of the RBM is strongly dependent on the diameter of the tube, and this can be demonstrated using a simple force constant model considering only nearest-neighbour interactions. Consider (Figure 6.6) a ring of n atoms joined by identical springs. The distance between adjacent atoms $c(t)$ oscillates about its equilibrium value c_0, and hence the tube

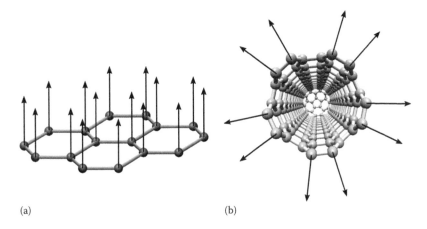

(a)   (b)

FIGURE 6.5
Eigenmode for zero-frequency out-of-plane phonon at the Γ point in graphene (a) and the non-zero frequency vibration this corresponds to in a SWCNT (b).

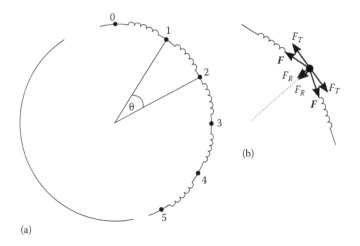

(a)

FIGURE 6.6
Nearest-neighbour force-constant model of radial breathing motion of a circular ring of n atoms (a). Forces exerted by nearest neighbours on an individual atom in the ring (b).

radius $r(t)$ also oscillates, about an equilibrium value of r_0. We assume that the equilibrium value of c_0 is independent of the number of atoms in the ring, and we can hence write

$$nc(t) = 2\pi r(t) \tag{6.1}$$

$$nc_0 = 2\pi r_0$$

When the ring of atoms is undergoing radial breathing motion, the restoring force F (Figure 6.6) is exerted along the axis of the interatomic bond and is given by $F = q(c(t) - c_0)$.

q is the spring constant of the interatomic bond. For each atom, the components of this force in the tangential direction F_T from the two adjacent bonds cancel out. We are considering an oscillation of the ring of atoms in which the tube radius changes, but there is no movement of atoms in the tangential direction and θ (as defined in Figure 6.6) is independent of the displacement. Hence, the restoring force we are interested in is the total force on each atom in the direction towards the centre of the tube. This is given by $2F_R$ where

$$F_R = |F|\sin\frac{\theta}{2} = \frac{qc_0}{2r_0}(c(t) - c_0) \qquad (6.2)$$

We can now define the displacement of the atom in the circumferential direction about its equilibrium position as $u(t) = r(t) - r_0$. Using Equation 6.1, we may also write $u(t) = ((n/2\pi)(c(t) - c_0))$. Hence, we can write down the equation of motion for the oscillation of the atom in the circumferential direction:

$$2F_R = -m\ddot{u}(t) = q\left(\frac{2\pi}{n}\right)^2 u(t) \qquad (6.3)$$

Inspection of Equation 6.3 shows that this is simple harmonic motion with a frequency ω which is dependent on the number n of atoms in the ring and hence on the radius r_0 of the ring. We can hence write

$$\omega \propto \frac{1}{r_0} \qquad (6.4)$$

As we shall see in the next chapter, this prediction agrees reasonably well with the results of more sophisticated theoretical models and also with available experimental data.

References

1. R Al-Jishi and G Dresselhaus, *Phys. Rev. B* **26**, 4514 (1982).

2. R Nicklow, N Wakabayashi and HG Smith, *Phys. Rev. B* **5**, 4951 (1972).

3. N Mounet and M Marzari, *Phys. Rev. B* **71**, 205214 (2005).

4. J Maultzsch, S Reich, C Thomsen, H Requardt and P Ordejón, *Phys. Rev. Lett.* **92**, 075501 (2004).

5. M Mohr et al., *Phys. Rev. B* **76**, 035439 (2007).

6. MB Walker, *Can. J. Phys.* **56**, 127 (1978).

7. JF Scott, *Ferroelectrics* **47**, 33 (1983).

8. AB Kuzmenko et al., *Phys. Rev. Lett.* **103**, 116804 (2009).

9. S Siebentritt, R Pues, K-H Rieder and AM Shikin, *Phys. Rev. B* **55**, 7927 (1997).

10. RA Jishi, L Venkataraman, MS Dresselhaus and G Dresselhaus, *Chem. Phys. Lett.* **209**, 77 (1993).

11. D Sánchez-Portal, E Artacho, JM Soler, A Rubio and P Ordejón, *Phys. Rev. B* **59**, 12678 (1999).

7 Raman Spectra of Graphene and Single-Walled Carbon Nanotubes (SWCNTs)

7.1 Overview

Raman spectroscopy has been utilized in the characterization of graphite since 1970 [1]. The technique was therefore ready to be used in the study of carbon nanotubes and graphene from a very early stage in the research on these materials. Shortly after the synthesis of multi-walled ([13] of Chapter 1) and single-walled ([14,15] of Chapter 1) carbon nanotubes in 1991 and 1993, Raman spectroscopy was utilized as a tool for characterization [2]. Raman spectroscopy has also been utilized in the characterization of graphene at an early stage [3] and has been widely used as a method to characterize graphene ever since [4]. In both cases (graphene and carbon nanotubes) the popularity of Raman spectroscopy stems from the fact that it is fast, cheap (relatively speaking), non-destructive, does not require the use of specific environmental conditions such as low temperature or vacuum and reveals a wealth of information about the samples studied.

The Raman spectra of graphene, all carbon nanotubes and, for that matter, graphite and diamond, reveal information about the presence of defects, strain, nature of the bonding (sp^2 or sp^3), number of layers in the case of graphene, and diameter and chirality in the case of SWCNTs. The primary disadvantage of Raman scattering as a method to characterize carbon materials is the fact that it is often an indirect measurement of the properties in which we are interested. A number of competing phenomena affect the Raman spectrum, and deconvolution of these to give accurate interpretation is very hard.

We will begin by understanding the Raman spectrum of pristine graphene then build on this understanding to examine the Raman spectra of bi-layer and few-layer graphene, graphite and SWCNTs. We will also examine the Raman spectra of defective graphene and graphite, and the use of Raman spectroscopy to characterize the degree of crystalline order in graphene and graphite samples.

In this review, we will consider only Raman scattering utilizing laser excitation not polarized along any specific axis. It is, however, worth noting that excitation with laser excitation polarized along a specific direction relative to the graphene, graphite or SWCNT crystal can enable even more information to be extracted from the technique. In the case of SWCNTs,

Raman scattering intensity is strongly attenuated when the incident light is polarized perpendicular to the axis of the tube [5]. The interested reader is referred to various publications dealing with this topic, for instance, references ([19] of Chapter 5 and [4,6]).

Readers not familiar with Raman scattering from solids may wish to read Appendix A before continuing, which is an introduction to this topic.

7.2   Principal peaks in the Raman spectrum of pristine mono-layer graphene

We can begin our detailed understanding of the Raman spectra of carbon materials by referring to the phonon dispersion relation of graphene in Figure 6.1. The conservation of momentum in Raman scattering ensures that only phonons at the centre of the Brillouin zone (the Γ point) can be created or annihilated in a first-order Raman scattering process unless another body is present and can receive a change in a momentum. This is the fundamental $k \approx 0$ selection rule (Appendix A). We therefore expect at most two peaks in the first-order Raman spectrum of graphene, at ca. 110 meV (885 cm^{-1}) and 195 meV (1569 cm^{-1}).

Out of these phonon modes, only the higher energy one is actually Raman active* and the Raman spectrum of pristine graphene (Figure 7.1) therefore shows a single intense peak from first-order Raman scattering at 1583 cm^{-1}, in reasonably good agreement with the ab initio prediction. This is the G peak.

The other intense peak in the Raman spectrum of graphene is the 2D† peak at 2860 cm^{-1}. This peak is the result of a second-order Raman scattering process in which two phonons are created. Since these phonons can have equal and opposite momentum, the conservation of momentum no longer dictates that these phonons are at the Γ point. The two phonons created are instead the eigenmode at ca. 1340 cm^{-1} at the K point ([19] of Chapter 5 and [4,7]) (the eigenmode labelled as (l) in Figures 6.1 and 6.2).

Generally, second-order Raman scattering is extremely weak compared to first-order Raman scattering. For instance, the second-order peak from diamond is roughly a factor of 50 less intense than the first-order peak.

However, in graphene the 2D peak is more intense than the (first-order) G peak. This is due to the fact that the 2D

---

* Not all phonons at the Γ point are Raman-active. Acoustic phonons are not Raman active and optical phonons are only Raman active when certain conditions are met. The interested reader is referred to Appendix A for discussion and examples on this point.

$\dagger$ Whilst "2D" is the standard notation in use, other notations have been used, particularly G' and D*. Readers may encounter these, particularly in older publications.

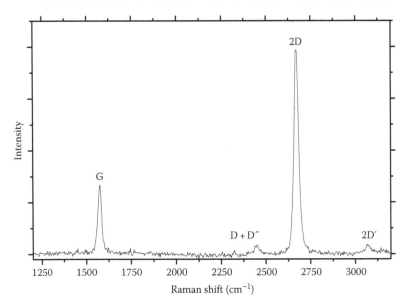

FIGURE 7.1
Raman spectrum of pristine mono-layer graphene with the G, D + D″, 2D and 2D′ peaks labelled.

peak is produced by a resonant Raman scattering process in which an electron is promoted from the valence band to the conduction band. This enhances the Raman scattering intensity drastically. Typically, Raman spectroscopy is performed using green or red laser excitation, so photons with an energy of approximately 1.5 – 2.5 eV. It is therefore only possible to promote an electron lying very close to the K point into the conduction band (see Figure 3.2). Figure 7.2 illustrates, on the electronic dispersion relation, the four stages in the scattering process responsible for the 2D peak in graphene. Since the electron exists in a real energy level both upon its initial excitation into the conduction band and after creation of the phonons, the process is referred to as double-resonant Raman scattering.

Given that, due to the creation of two phonons, the $k \approx 0$ selection rule is relaxed, one might reasonably ask: Why is the specific phonon eigenmode at the K point the only phonon which produces a strong second-order Raman peak? The answer to this question also lies in the double-resonant nature of the Raman scattering process producing the 2D peak. This process – we have already established – is necessary to produce a second-order Raman peak with significant intensity. The electron excited in this process has to be excited close to the K point as this is the only location in the Brillouin zone where the energy of a visible laser photon is sufficient to excite an electron from the valence band to the conduction band. Thus, the phonons involved in this process must have the right wavevector to scatter the electron from

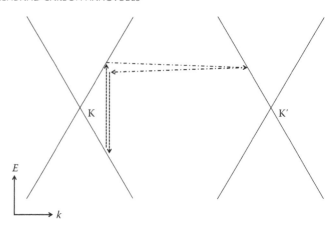

FIGURE 7.2
Double-resonant Raman scattering process responsible for the observation of the 2D Raman peak in mono-layer graphene. The dotted lines correspond to electron–photon scattering, and the dot-dash lines to electron–phonon scattering. The wavevector plotted along the x-axis corresponds to a movement through **k**-space along the edge of the Brillouin zone from the K point, through the M point to the other (inequivalent) K point, labelled here as K′.

one K point to the other K point. Since the area ΓKK is an equilateral triangle in reciprocal space, this means that the phonons excited in the process shown in Figure 7.2 must also be close to the K point.

The eigenmode specified earlier is the only optical phonon meeting the requirements in terms of polarizability (see Appendix A) to be Raman active and exhibiting a significant density of states* (flat dispersion relation) at the K point.

7.2.1   Energy-dispersive nature of the 2D peak

In a conventional non-resonant first-order Raman scattering process, the energy and frequency of the Raman peak do not shift as a function of excitation laser energy. This is because the peak position is the energy of the zone-centre phonon which produces the peak, which is independent of the excitation laser photon energy.

However, the double-resonant nature of the Raman scattering process that produces the 2D peak results in the highly unusual characteristic that the Raman peak position does shift as a function of laser excitation energy. Referring to

---

* A large phonon density of states is necessary to observe an intense Raman peak. The easiest way to understand this is to consider, as an alternative to our classical treatment in Appendix A, the Raman scattering as a transition between two quantum states. Thus the transition rate is determined by Fermi's golden rule and is directly proportional to the density of states available to the system after the scattering process.

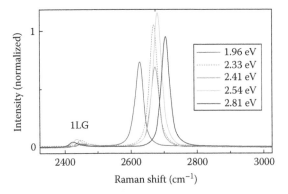

FIGURE 7.3
Raman 2D peak of pristine mono-layer graphene collected at a variety of different excitation laser energies. (Reprinted by permission from Macmillan Publishers Ltd. *Sci. Rep.*, An Nguyen, T., Lee, J.-U., Yoon, D. and Cheong, H., 4, 4630, copyright 2014.)

Figure 7.2, we can understand why this is the case. If the excitation laser energy is changed, the electron is excited from the valence band to the conduction band at a slightly different point in reciprocal space. Therefore, the wavevector and energy of the phonon required to scatter the electron into the conduction band above the other K point changes slightly. So a different phonon is excited in the Raman scattering process. Figure 7.3 illustrates the shift in Raman 2D peak position as a function of laser excitation energy as a result [8].

7.2.2   Other Raman peaks in pristine graphene

In addition to the intense G and 2D peaks, there are two weaker peaks also present in the Raman spectrum of pristine mono-layer graphene (Figure 7.1). These are at ca. 2450 cm^{-1} (the D + D″ peak) and at ca. 3250 cm^{-1} (the 2D′ peak). Both of these peaks originate from second-order Raman scattering processes in which two phonons with equal and opposite wavevector are created.

The phonons contributing to the D + D″ peak are the same phonon as that which causes the 2D peak, and the eigenmode at ca. 1200 cm^{-1} at the K point (the phonon in the LA branch which originates at $E = 0$ at the Γ point). Whilst these phonons have different energy, they can be excited with equal and opposite wavevector since they both lie at the K point. Hence the conservation of momentum can be satisfied during the resonant Raman scattering process.

The 2D′ peak originates from a double-resonant Raman scattering process similar to the 2D peak, except that after excitation into the conduction band the electron scatters into another location in the same Dirac cone (Figure 7.4). Hence, the

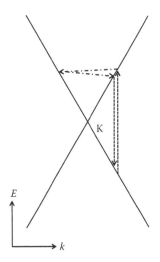

FIGURE 7.4
Double-resonant Raman scattering process responsible for the 2D′ peak in pristine mono-layer graphene. The dotted lines correspond to electron–photon scattering, and the dot-dash lines to electron–phonon scattering.

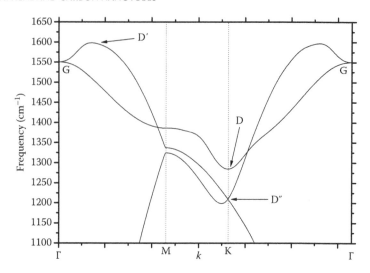

FIGURE 7.5
Detail of the high-energy section of the graphene phonon dispersion relation (from Figure 6.1), with phonons responsible for the D + D″, 2D and 2D′ peaks labelled.

phonon created in this process has a much smaller wavevector than that responsible for the 2D peak. It is the optical phonon at ca. 1600 cm^{-1} between the Γ point and the M point.

Figure 7.5 summarizes and clarifies our understanding of which phonon(s) lead to which Raman peaks. Figure 7.5 details the higher energy part of the graphene phonon dispersion relation (Figure 6.1) with phonon modes D, D′ and D″ labelled. Note that phonons with a range of wavevector could contribute to the 2D′ peak, but that we have labelled the point on the dispersion relation where the phonon density of states (as a function of energy) will be largest as we expect the Raman peak intensity to be dominated by phonons from this point.

7.3   Raman spectra of pristine bi-layer and few-layer graphene, and pristine Bernal stacked graphite

The Raman spectra of pristine bi-layer, few-layer graphene and pristine Bernal (ABAB) stacked graphite exhibit just two intense peaks (Figure 7.6): the G peak and the 2D peak, just as for mono-layer graphene. The G peak originates from a Raman scattering process identical to that which produces the G peak in mono-layer graphene.* The 2D peak, however, undergoes

---

* The peak position shifts very weakly as a function of the number of layers, from 1580 cm^{-1} in graphite to 1583 cm^{-1} in mono-layer graphene. The shift is at least partially due to the interaction with the substrate [9] and is too small for reliable characterization of graphene samples.

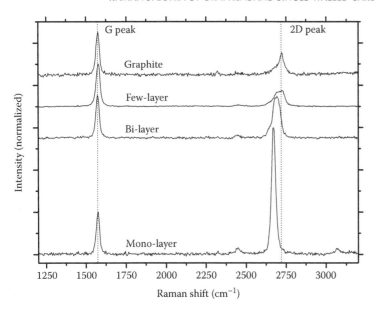

FIGURE 7.6
Raman spectra of mono-layer graphene, bi-layer graphene, few-layer graphene and graphite. The first-order G peak and second-order D peak (the 2D peak) are labelled. Intensities are normalized so that the G peak intensity is the same in each spectrum.

substantial changes in position, intensity and lineshape depending on the thickness of the sample. As a result, Raman spectroscopy is a fast and non-destructive method to distinguish mono-layer graphene, bi-layer graphene, few-layer graphene and graphite.

As shown in Figure 7.2, the double-resonant Raman scattering process responsible for the observation of the 2D peak in graphene commences with the excitation of an electron from the valence band to the conduction band. The electron is then scattered to the conduction band close to the other (inequivalent) K point, and then back again (with the change in wavevector in each scattering being the wavenumber of the phonon created). Considering that the energy provided to the electron for the initial excitation is fixed (the excitation laser energy), we can see that if the electronic dispersion relation changes, the wavenumber of the phonon for which conservation of momentum in this process is satisfied also changes.

Therefore, the change in the electronic dispersion relation as the thickness of a graphene layer changes alters the characteristics of the 2D peak. Also, the change from the very simple dispersion relation of mono-layer graphene to the more complex dispersion relation for graphite allows several different excitation processes to occur for any given

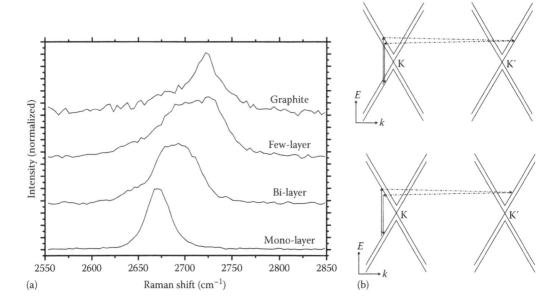

FIGURE 7.7
(a) 2D Raman peaks of mono-layer, bi-layer and few-layer graphene, and graphite.
(b) Double-resonant Raman scattering processes responsible for the two most intense
components of this peak in bi-layer graphene, illustrated on the bi-layer graphene electronic dispersion relation.

laser wavelength – each of which results in the excitation of a
phonon with slightly different energy and an additional component to the 2D peak.

Figure 7.7 shows the 2D Raman peak in more detail for
mono-layer, bi-layer and few-layer graphene, and also for
graphite. The two different double-resonant Raman scattering processes leading to the two most intense components
of this peak in bi-layer graphene are shown. The electronic
dispersion relation of bi-layer graphene has two branches
near the K point, leading to four different possible double-
resonant Raman scattering processes and four components
in the 2D peak.

In addition to the change to a more complex lineshape, the
intensity of the 2D peak changes relative to that of the G peak
as the thickness of the graphene sample changes. The spectra
shown in Figure 7.6 are normalized so that the intensity of the
G peak is the same in each case. As we can see, the intensity
ratio of the G and 2D peaks changes enormously as the number of layers (thickness) of the graphene sample changes. For
a mono-layer, the peak intensity of the 2D peak is three times
that of the G peak, whilst at the other extreme in graphite the
2D peak is significantly less intense than the G peak.

Finally, there is a single Raman peak present in bi-layer
graphene, few-layer graphene and graphite but not in mono-
layer graphene. This is a peak due to the excitation of shearing

motion of two graphene planes. This results in a weak peak (referred to as the C peak) at extremely low wavenumber (42 cm^{-1}) [10–12]. The peak lies too close to the intense signal from the elastically scattered laser excitation light to be detected by a conventional single grating Raman spectrometer (though it is worth noting that the peak does shift to higher wavenumber under pressure [12]). Hence the peak is not present in the spectra in Figures 7.1 and 7.6 and its absence does not constitute convincing evidence for the presence of mono-layer graphene. The low energy of this phonon mode does ensure that a substantial anti-Stokes peak is observed even at ambient temperature [10] (see Appendix A) and allows potential applications in the accurate measurement of the temperature of all graphene samples except for mono-layer samples.

7.4   Raman spectra of pristine graphite with different stacking arrangements

In the previous section, we considered the evolution of the Raman spectrum of graphene as sample thickness is increased from a mono-layer of graphene to bulk graphite, for the case of Bernal stacking. Bernal (ABAB) stacking of the graphene layers is the thermodynamically stable and most common stacking arrangement found in nature, but other stacking arrangements do exist ([11] of Chapter 1 and [4]). The electronic dispersion relation (and thus the Raman spectrum) is dependent on the stacking arrangement. In particular, the 2D Raman peak of turbostratic graphite (graphite in which successive layers are oriented randomly with respect to each other) consists of just a single Lorentzian [3,4,13]. Therefore, a 2D peak consisting of just a single Lorentzian is not adequate as the sole evidence for the presence of mono-layer graphene.

Fortunately, the 2D peak of turbostratic graphite is substantially broader than that of mono-layer graphene and lies at substantially higher wavenumber (~20 cm^{-1} higher). These features allow it to be distinguished from the 2D peak in graphene.

7.5   Characterization of pristine graphene samples using Raman spectroscopy

It is accepted that using Raman spectroscopy it is possible to place a pristine graphene/graphite sample into one of the three categories:

1. Mono-layer graphene

2. Bi-layer graphene

3. Few-layer graphene or graphite

This is achieved by measuring the intensity ratio of the G and 2D peaks, combined with the lineshape and position of the 2D peak. Note, however, that it is not possible to quantitatively measure the thickness of a few-layer graphene sample using Raman spectroscopy, because, as a fourth, fifth or sixth layer is added to a graphene sample the changes in the 2D peak are too subtle to use to accurately characterize how many layers are present. All we can conclude with certainty is that the sample has more than two layers. The characterization of graphene samples in this manner utilizing Raman spectroscopy has been verified using the more direct, but less convenient, method of electron diffraction [3].

7.6   Raman spectrum of diamond

It is useful to compare the Raman spectrum of pristine sp^2-bonded carbon (graphene) with pristine sp^3-bonded carbon (diamond). This is principally because, when we chemically modify graphene, we are causing a (partial) conversion to sp^3-bonding. The Raman spectra of chemically modified graphene can be understood in this context. The first point to note is that the Raman scattering cross-section of sp^3-bonded carbon is ~50× lower than that of sp^2-bonded carbon under excitation with visible light [14] (though not under ultraviolet excitation [4]). In practice the first-order Raman spectrum of diamond is orders of magnitude more intense than that from graphite (in fact, it is so intense that it was observed in the 1930s prior to the development of lasers [15]). This is, however, solely due to the fact that since diamond is transparent there are many more atoms contributing to the Raman spectrum.*

Unlike graphite, graphene and SWCNTs, in which the conduction and valence bands meet or come close at the K point, diamond is a wide bandgap insulator. There is thus no possibility for resonant Raman scattering to take place and the only intense feature in the first-order Raman spectrum of diamond is the peak at 1332 cm^{-1}, originating from the only phonon mode at the Γ point with a non-zero energy and frequency ([3] of Chapter 6). This mode involves motion of atoms with alternate atoms out of phase with each other. Since the Raman scattering process is non-resonant, the second-order Raman peak for this mode is far less intense than the first-order peak (Figure 7.8).

---

* Graphite is sufficiently opaque that visible photons penetrate less than 100 atomic layers.

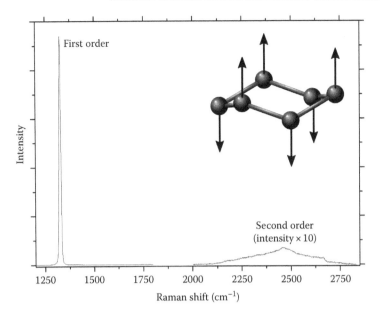

FIGURE 7.8
(Main figure) Raman spectrum of diamond, showing first- and second-order peaks. (Inset) Raman-active phonon eigenmode in diamond.

7.7 Raman spectra of defective graphene and graphite

The diagnosis of the presence of defects in graphene (and graphite) is one of the principal applications of Raman scattering in this field. We can divide our discussion of the Raman spectra of defective graphene and graphite into two parts: the spectra of samples with small concentrations of defects and the spectra of samples with large concentrations of defects.

7.7.1 Small defect concentrations

The principal effect of small concentrations of defects is to enable several additional double-resonant Raman scattering processes in which just one phonon not from the zone centre is created. Earlier we discussed the double-resonant Raman scattering process leading to the observation of the 2D peak in pristine graphene and graphite. The creation of two phonons ensured the conservation of momentum in this process despite the phonons not being from the zone centre. In defective graphene and graphite, it is possible for the conservation of momentum to be satisfied during a process in which the creation of one of these phonons is replaced by an event in which the electron scatters from a defect in the graphene lattice.

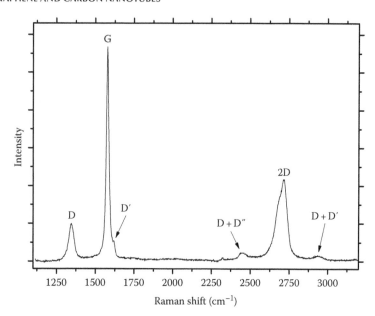

FIGURE 7.9
Raman spectrum of graphite with a small concentration of defects. Compared to the spectrum of pristine graphene (Figure 7.1) additional peaks D, D′ and D + D′ appear.

Therefore, the Raman spectrum of graphite with a small concentration of defects (Figure 7.9) exhibits additional features compared to its pristine graphene or graphite counterpart (Figures 7.1 and 7.6). The strongest of these originates from a double-resonant Raman scattering process in which a single phonon at the ca. 1340 cm^{-1} eigenmode at the K point is created. In this case, the scattering process is the same as that shown in Figure 7.2 except that one of the processes in which the electron moves from one Dirac cone to the other involves the electron scattering from a defect instead of the creation of a phonon. The resulting Raman peak is called the D peak. In addition, a weaker peak appears known as the D′ peak. This stems from a similar process, except that a single phonon at ca. 1600 cm^{-1} from the region between the Γ point and the M point is created. There is also an additional second-order Raman peak known as the D + D′ peak in which a D phonon and a D′ phonon are created. This process is not permitted in pristine graphene because these phonons have different wavevector.

7.7.2   Large defect concentrations

In our discussion in the previous section, we accounted for the presence of defects only by allowing electrons participating in the resonant Raman scattering process to scatter from them. This allows several additional Raman scattering processes which do not satisfy the fundamental Raman selection rule in pristine graphene.

However, if we allow the concentration of defects to increase, then eventually there will be enough defects to cause considerable disruption to the periodic structure of the graphene lattice. Therefore, the reciprocal lattice will also be disrupted and this will affect the phonon and electronic dispersion relations.

This process results in further qualitative changes to the Raman spectrum. The intense second-order peaks disappear due to the disruption to the electron dispersion relation and hence lack of a resonant Raman scattering process. The only Raman peaks observed are weak and broad D and G peaks (Figure 7.10). The D peak corresponds to a breathing mode of a hexagonal aromatic ring, just as in pristine graphene

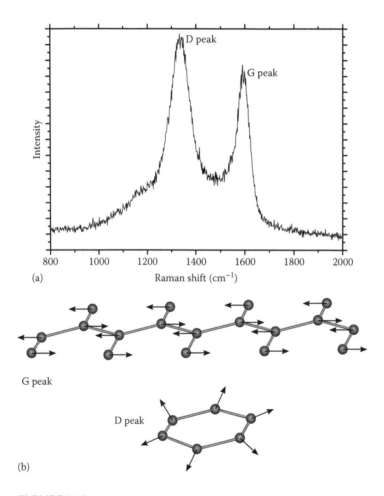

(a)

(b)

FIGURE 7.10

(a) Raman spectrum of highly disordered carbon (activated charcoal). (b) Vibrations corresponding to the G and D peaks in a disordered carbon sample.

(Figures 6.1 and 6.2), and hence only requires structural order to extend to the level of single rings for its activation.

The G peak is an allowed zone-centre phonon in pristine graphene and, at the other extreme (amorphous carbon), does not even require the presence of aromatic rings to be observed. Figure 7.10 shows the vibration corresponding to the D peak in a single aromatic ring and the vibration corresponding to the G peak in a chain of sp^2-bonded carbon atoms. Indeed, the G peak is observed in molecules consisting of chains of carbon atoms [7].

The ratio of the intensities of the D and G peaks (I_D/I_G) can be used to estimate the concentration of defects in graphene both in the regime described in the previous section (small defect concentration) and the regime described in this section (a sample with significant amorphous character). This observation was first made by Tuinstra and Koenig in 1970 [1]. It was later developed by Sato et al. in 2006 [16] and most recently by Lucchese et al. in 2010 [17].

Figure 7.11 plots the correlation between the average spacing between defects and the I_D/I_G intensity ratio observed in the study by Lucchese et al. At one extreme (nearly pristine graphene, large spacing between defects), the I_D/I_G ratio converges towards zero due to the disappearance of the D peak. As the spacing between defects is gradually reduced the D peak increases in intensity until it is over 3× the intensity of the G peak (the regime discussed in Section 7.7.1), then decreases in intensity as the sample becomes increasingly amorphous (current section).

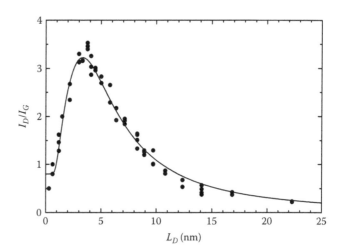

FIGURE 7.11

Graph of experimentally observed I_D/I_G intensity ratio in the graphene Raman spectrum as a function of the average spacing between adjacent defects L_D. (Reprinted from *Carbon*, 48, Lucchese, M.M. et al., 1592, Copyright 2010, with permission from Elsevier.)

Whilst the correlation observed in Reference 17 is very clear, it is important to note that this is one study utilizing one variety of graphene sample, one laser excitation energy and a single method to create the defects. The exact relationship observed between the defect spacing and the I_D/I_G ratio cannot be universally applied as the results must depend on many different variables. For instance, the presence of large numbers of defects may induce some sp^3 hybridization, for which the Raman scattering cross section varies differently as a function of excitation laser energy compared to sp^2 hybridized carbon. In the regime with fewer defects, we would expect different kinds of defects to interact differently with the electrons, hence affecting the Raman scattering cross section for the double-resonant Raman scattering process. So the specific correlation shown in Figure 7.11 applies only to the exact experimental conditions in Reference 17. The reader may also wish to refer to Appendix A, in which Raman spectra of graphitic materials with a range of different defect densities are shown (Figure A.2).

7.8   Raman spectra of SWCNTs

The Raman spectra of SWCNT differ from the spectra of other graphene and graphite-like materials in two key ways. Firstly, all observed Raman peaks stem from resonant Raman scattering processes. This is a result of the existence of direct extremal bandgaps in the electronic dispersion relation of SWCNTs (Chapter 5) that do not exist in the graphene dispersion relation. As a result, the Raman spectrum of SWCNTs is much more intense than that of graphene or graphite. It seems remarkable that we can observe a Raman spectrum from a single atomic layer of graphene, yet the number of carbon atoms illuminated by the ~1 μm diameter laser beam when we take the spectrum of graphene is enormous compared to the number in the beam when we take the spectrum of a single SWCNT with ~1 nm diameter. Yet it is possible to observe a resonant Raman spectrum from a single isolated SWCNT ([6,19] of Chapter 5 and [18]).

Secondly, changes to the phonon dispersion relation as a result of rolling the graphene sheet up into a tube are reflected in the Raman spectrum. As discussed in Section 6.4.2, the out-of-plane phonon with zero energy and zero wavevector at the Γ point in graphene becomes a phonon with non-zero energy in SWCNTs. This phonon is Raman active (the radial breathing mode or RBM) and its frequency can be used to determine the diameter of an SWCNT. The high-energy phonon responsible for the graphene and graphite G peak splits into two modes in the SWCNT, depending on whether the vibration is parallel or perpendicular to the SWCNT axis.

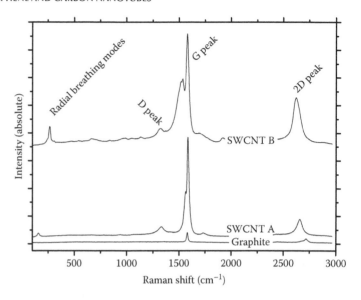

FIGURE 7.12

Raman spectra of SWCNTs and pristine graphite. SWCNT sample A is produced using the Carbolex® process, and sample B is produced using the HiPCO® process. In all cases the radial breathing modes, D peak, G peak and 2D peak are shown. The much weaker nature of the Raman scattering from graphite is demonstrated: The spectra were taken consecutively under the same experimental conditions and have not been normalized.

Figure 7.12 shows the Raman spectra of bundled SWCNTs produced in two different growth processes to give an overview of the different peaks and their relative intensities. In order of increasing frequency, we begin with the RBMs at 100–300 cm^{-1}. These relatively weak modes correspond to the breathing vibrations of different diameter SWCNTs. At higher frequency is the D peak at ca. 1350 cm^{-1}, which originates from a defect-induced double-resonant Raman scattering process similar to the equivalent peak in graphite. In SWCNTs however, the curvature of the tube also plays a role in breaking the translational symmetry of the lattice and allowing this process to occur. At 1550–1600 cm^{-1} is the G peak which splits in SWCNTs for symmetry reasons, and at ca. 2650 cm^{-1} is the 2D peak which originates from a double-resonant Raman scattering process not requiring the presence of defects, similar to the equivalent peak in graphite.

7.8.1   Resonant nature of Raman scattering from SWCNTs

We will begin by making some general points regarding the resonant Raman scattering process in graphene and SWCNTs. Figure 7.13 shows the electronic dispersion relation

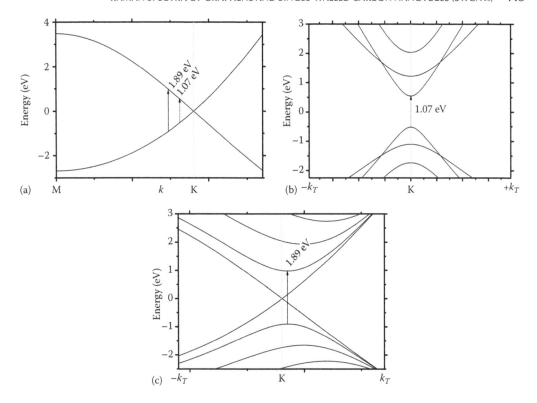

FIGURE 7.13

Electronic dispersion relation of graphene (a), the (10,0) SWCNT (b) and (10,10) SWCNT (c) in the vicinity of the Fermi level. Excitation possible utilizing 1.89 eV (657 nm) laser radiation is marked using the solid line on the graphene and (10,10) dispersion relations, while excitation possible utilizing 1.07 eV (1160 nm) laser radiation is marked using the dotted line on the graphene and (10,0) dispersion relations. The K point is marked with the vertical grey line. Excitonic effects are omitted for clarity.

of graphene (predicted using the tight binding procedure described in Chapter 3), as well as the electronic dispersion relation of the (10,0) and (10,10) SWCNTs predicted using the zone-folding procedure in Chapter 5. We concentrate on the region close to the Fermi level, where electrons may be excited from the valence band to the conduction band. In the case of the (10,0) SWCNT near-infrared (1.1 eV) laser excitation can be utilized to excite an electron across an extremal bandgap; that is, there is a flat dispersion relation at this point and a large density of states (van Hove singularity) available for electrons before and after the transition. According to Fermi's golden rule, we will therefore observe a large transition rate. In the case of the (10,10) SWCNT, this can be achieved using green laser excitation (1.9 eV), whilst in the case of graphene both these photon energies excite electrons from the linear region of the dispersion relation close to the K point where the density of states available before and after excitation is much lower.

For this reason, all peaks observed experimentally in the Raman spectra of SWCNTs originate from resonant Raman scattering processes. The experimentally observed spectra of SWCNTs are much more intense than those from graphene or graphite (Figure 7.12) and the very fact that a spectrum observed provides information about the SWCNT sample: It tells us that the excitation laser energy utilized lies close to one of the extremal electronic bandgaps in some of the SWCNTs in the sample.

The resonant Raman scattering process responsible for the D and 2D peaks in SWCNTs is a double-resonant process similar to that responsible for the equivalent peaks in graphene. However, the other Raman peaks observed in SWCNTs (the radial breathing mode, Section 7.8.2, and the G peak, Section 7.8.3) all originate from the excitation of single zone-centre phonons. The resonant Raman scattering process responsible for these peaks is believed to be different to that responsible for the 2D peak – it is referred to as single-resonant Raman scattering, although the potential for a double-resonant Raman scattering process to produce the G peak has also been discussed ([19] of Chapter 5).

Figure 7.14 illustrates, for the case of the (10,10) SWCNT discussed in Chapter 5, examples of the single-resonant Raman scattering process responsible for the observation of the radial breathing mode (Section 7.8.2) and G peak (Section 7.8.3) in

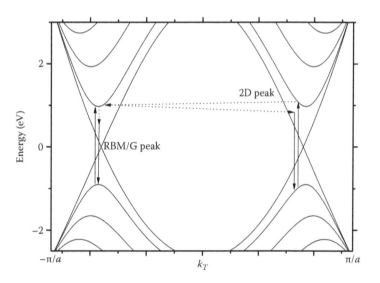

FIGURE 7.14

Calculated electronic dispersion relation of the (10,10) SWCNT with single-resonant (responsible for the RBM and G peak) and double-resonant (responsible for the D and 2D peak) Raman scattering processes marked. Excitonic effects are omitted for clarity.

SWCNTs. These are zone-centre phonons with zero wavenumber so it is not necessary for the electron to receive a change in momentum when the phonon is created. Instead, the electron drops straight back into the hole it left behind in the valence band, creating a phonon and emitting a photon with energy slightly lower than the incident laser photon.

Figure 7.14 also illustrates the double-resonant Raman scattering process responsible for the observation of the D and 2D peaks in SWCNTs. The process is similar to that responsible for the observation of these peaks in graphene in that the electron moves from one K point to the other each time a phonon is created, but different in that the initial excitation from the valence band to the conduction band does not have to be via the Dirac cones that meet at the K point. Instead, it can be via the (approximately) parabolic bands which exhibit an extremal bandgap close to the K point. Due to the parabolic nature of the bands the density of states available for the electrons near the extremal bandgap between the parabolic bands is far greater than that in the linearly dispersive bands that meet at the K point.

In Section 7.2.1, we discussed the energy-dispersive nature of the double-resonant Raman scattering process leading to the observation of the 2D peak in graphene. In this case, whilst the 2D peak position shifts as a function of the laser excitation energy, it can in all cases be observed. However, in SWCNTs the single-resonant (radial breathing mode and G peaks) and double-resonant (D and 2D peaks) Raman scattering processes which lead to the observation of Raman spectra far more intense than those from graphene and graphite are reliant on the excitation laser wavelength coinciding with the energy gap between two of the van Hove singularities. Hence the resonant Raman scattering process can only take place in a specific SWCNT if the excitation laser energy coincides with the energy gap between two of the van Hove singularities. Since the electron cannot receive a significant change in momentum (wavevector) during this process, it must be excited from a van Hove singularity in the valence band to the singularity in the same branch* of the dispersion relation in the conduction band. The energies of the allowed excitation processes are labelled according to their proximity to the K point, with E_{11} being the closest to the K point and so on. Figure 7.15 illustrates E_{11}, E_{22} and E_{33} for the case of the (10,0) SWCNT. These are the same energy gaps as those we labelled in Figure 5.8.

---

* Here, we mean the branch of the conduction band with the same quantized value of wavevector around the circumference of the SWCNT, and the same value of μ in Equation 5.6 and subsequent equations.

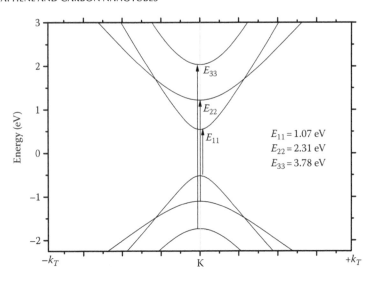

$E_{11} = 1.07$ eV
$E_{22} = 2.31$ eV
$E_{33} = 3.78$ eV

FIGURE 7.15
The energies of the first three allowed excitation processes between van Hove singularities in the (10,0) SWCNT. Excitonic effects are omitted for clarity.

7.8.2   Radial breathing mode (RBM)

In Section 6.4.2, we discussed the vibration corresponding to a breathing motion of the SWCNT in which the entire tube expands and contracts. Such a vibration does not exist in graphene or graphite. Using a simple classical ball-on-a-spring model of an isolated SWCNT, we predicted that the frequency ω of this vibration would be inversely proportional to the SWCNT diameter d_t (Equation 6.4, restated in the following as Equation 7.1) in which we use the notation ω_i to denote the frequency for an isolated tube, and A is a constant. This vibration is Raman active and is known as the radial breathing mode (RBM):

$$\omega_i\left(d_t\right) = \frac{A}{d_t} \tag{7.1}$$

Experimental measurements of the RBM frequency agree reasonably well with the simple correlation between frequency and diameter that we proposed in Equation 7.1. Where the $\omega(d_t)$ relation deviates from that dictated by Equation 7.1 the frequency is always higher than we would expect. This can be attributed to the SWCNT encountering resistance from its environment as it tries to expand, which increases the effective restoring force and spring constant for the vibrational mode (Equation 6.2 onwards).

Empirically, three different $\omega(d_t)$ relations have been found to provide good fits to different sets of experimental data – the relation in Equation 7.1 that we obtained in Section 6.4.2

by considering a classical model of an isolated SWCNT, the addition of a constant B to the RBM frequency predicted for an isolated tube (Equation 7.2) and the more complex relation resulting from considering a classical model of an SWCNT vibrating under an external pressure which increases as it expands (Equation 7.3, where C is also a constant [19]):

$$\omega(d_t) = \frac{A}{d_t} + B = \omega_i + B \qquad (7.2)$$

$$\omega(d_t) = \frac{A'}{d_t}\sqrt{1 + Cd_t^2} = \omega_i + \frac{A'}{d_t}\left[\sqrt{1 + Cd_t^2} - 1\right] \qquad (7.3)$$

A systematic study of the RBM frequency of isolated tubes grown protruding vertically upwards from a silicon substrate ([15] of Chapter 5 and [20]) produced results that were fitted well by Equation 7.2 over a large diameter range (1–4 nm, Figure 7.16), with a value of A closely in line with theory ([15] of Chapter 5) and an extremely small value of B ($A = 227.0$ and $B = 0.3$ cm^{-1}). The observed $\omega(d_t)$ dependence was therefore close to the simple inverse relationship predicted in Equations 6.4 and 7.1. This is what we expect given the isolated nature of the SWCNTs studied in this work.

A number of studies of SWCNTs in different environments produced data fitted well by Equation 7.2, with values for A in the region of 200–250 cm^{-1} and values for B of up to 30 cm^{-1}

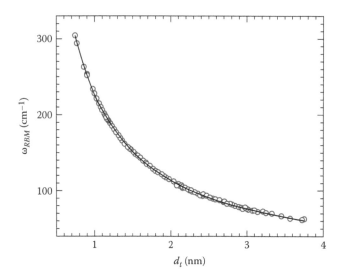

FIGURE 7.16

Experimentally observed RBM frequency ω_{RBM} of isolated SWCNTs protruding vertically from a silicon substrate plotted as a function of tube diameter d_t. (Reprinted from *Physica E*, 42, Araujo, P.T., Pesce, P.B.C., Dresselhaus, M.S., Sato, K., Saito, R. and Jorio, A., 1251, Copyright 2010, with permission from Elsevier.)

([15] of Chapter 5 and [21]), or Equation 7.3 [20]. Equation 7.1 only provides a satisfactory fit to the data for SWCNTs which are nearly isolated.

Considering the earlier results, it is hopefully self-evident that it is possible to use a measurement of the SWCNT RBM frequency to measure the diameter of the tube. However, a number of different $\omega(d_t)$ relations are published in the literature (here we have reviewed only a small proportion of the published data). This is a result of the effect of different environments (for instance, different substrates and solvents) on the RBM frequency. Therefore, if we are using a measurement of RBM frequency to determine the diameter of SWCNTs and accurate results are required, it is essential to choose a $\omega(d_t)$ relation from the literature, which is valid for the environment of the SWCNTs being studied.

Since the RBM does not exist in graphite or graphene, one could be tempted to use its observation to prove the presence of SWCNTs in an unknown sample. In practice this is unwise since a diverse range of different materials also have Raman peaks in the frequency range where the RBM is typically found (100–300 cm^{-1}). In practice, to prove the presence of SWCNTs in a sample, it is also necessary to observe the G peak. Unlike graphene or graphite, in which the G peak at ambient conditions can be fitted by a single Lorentzian, in the SWCNT the peak splits for symmetry reasons. This phenomenon, discussed in the next section, provides convincing evidence for the presence of SWCNTs in a sample when combined with the observation of the RBM.

7.8.3   G peak

The G peak in graphene and graphite consists of a single Lorentzian peak at ca. 1580 cm^{-1}, assigned to the zone-centre optical phonon with E_{2g} symmetry excited in a non-resonant Raman scattering process. In SWCNTs, by contrast, the G peak splits into several components often with differing lineshapes. All Raman-active modes contributing to the G peak are believed to be excited in single-resonant Raman scattering processes. The G peak characteristics provide information about the diameter and metallic or semiconducting character of the SWCNT. A comprehensive treatment of the G peak would be beyond the scope of this introductory text (the reader is referred to more detailed works on this topic ([19] of Chapter 5 and [22–24])). Instead, we will list a few simple rules to give the reader a basic knowledge of the topic.

Firstly, the G peak corresponds to a vibration in which all the atoms vibrate out of phase along the same axis (see eigenmode at point B in Figures 6.1 and 6.2). In graphene or graphite, such a vibration (at the Γ point) can occur along three different directions in the graphene sheet and have the same energy. In SWCNTs, however, the energy of the vibration depends on its orientation relative to the tube axis. The need for quantization of the component of the phonon wavevector in the direction around the circumference of the SWCNT (the zone-folding procedure, as discussed in Chapter 6) causes further splitting in the allowed phonon modes at the Γ point. The experimentally observed G peak in SWCNTs is usually made up of two separate peaks (G+ at higher frequency and G− at lower frequency), corresponding to atomic displacements in-plane along the tube axis (LO) or perpendicular to the tube axis (TO).* The TO can correspond to the G+ peak and the LO can correspond to the G− peak, or vice versa, depending on the metallic/semiconducting nature and chirality of the SWCNT concerned [22,24].

Secondly, the magnitude of the splitting between the G+ and G− peaks increases with decreasing tube diameter ([12] of Chapter 1, [19] of Chapter 5 and [22,24]). Figure 7.17 illustrates this trend with Raman spectra from graphite (in which a single peak is observed), and bundled SWCNTs produced using the Carbolex® (mean diameter 1.4 nm) and HiPCO® (mean diameter 1.1 nm) processes. The smaller average diameter of

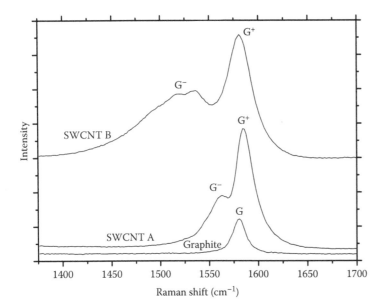

FIGURE 7.17

Raman G peak of graphite (bottom curve), Carbolex® SWCNTs (A, middle curve) and HiPCO® SWCNTs (B, top curve) illustrating increasing splitting between G+ and G− peaks with decreasing tube diameter. The intensity of the graphite spectrum is multiplied by 3 for clarity.

---

* Note that this is different to describing a phonon travelling along the SWCNT or around the circumference of the SWCNT. Since we are exciting zone-centre ($k \approx 0$) phonons only, it is not correct to assign a direction of travel to the phonon.

the HiPCO tubes results in a larger splitting between the G^+ and G^- peaks.

Thirdly, the lineshape and relative intensity of the G^+ and G^- components varies strongly according to the exact (n,m) assignment. In particular, all armchair (n,n) SWCNTs exhibit only a narrow G^+ peak [23]. All zigzag $(n,0)$ SWCNTs exhibit just one peak, a broad G^- peak in metallic zigzag tubes and a narrow G^+ peak in semiconducting zigzag tubes [23]. Non-armchair metallic SWCNTs exhibit both G^+ and G^- peaks. The G^- peak exhibits a highly asymmetric Breit–Wigner–Fano lineshape instead of the usual Lorentzian [23].

7.8.4   D and 2D peaks

A D peak and a 2D peak are observed in the Raman spectrum of SWCNTs. The D peak occurs due to a double-resonant Raman scattering process (Figure 7.14) which is different to that responsible for the graphene and graphite D peaks in one respect: The excitation of the electron from the valence band to the conduction band takes place across the extremal bandgap instead of from the linear section of the dispersion relation close to the K point. This results in a larger density of states for electrons to be excited from and scatter into, due to the existence of the van Hove singularities. Once excited into the conduction band the electron scatters to the conduction band above the other K point, then back again. Just as in the double-resonant Raman scattering process in graphene and graphite, one scattering event is due to the creation of a phonon and one is due to scattering from a defect. In the case of the 2D peak, both scattering events are due to creation of phonons. Hence, in good-quality SWCNT samples the 2D peak is much more intense than the D peak (Figure 7.12).

Whilst in graphene and graphite the D peak is excited due to the presence of edges or defects, in SWCNTs the curvature of the tube and partial conversion to sp^3 hybridization also plays a role [25]. However, in SWCNTs produced using modern methods (for instance, references [26,27]), the D peak frequently has a vanishingly small intensity relative to the G peak.

Both the D and 2D peaks are energy dispersive (i.e. the peak position shifts as a function of the laser excitation energy), but (unlike the case of mono-layer graphene) they have several different components due to the symmetry-breaking effect of the tube curvature and possibility to excite the electron across several different extremal bandgaps.

7.8.5 (n,m) Assignment in SWCNTs from Raman spectroscopy

It is possible to uniquely assign (n,m) indices to an SWCNT using Raman spectroscopy. However, great care is required with this procedure as there are many potential pitfalls that can produce erroneous results. Here we will outline the procedure and the potential pitfalls. To begin with, it is essential to ensure that the Raman spectrum originates from a single SWCNT. In some cases, it is possible to prove that an observed Raman spectrum originates from more than one SWCNT (for instance, if more than one separate RBM peak is observed), but it is not possible to prove that an observed Raman spectrum originates from a single SWCNT.

In a Raman spectroscopy experiment, the minimum area illuminated by the excitation laser beam (from which the spectrum is collected) is rather large on an atomic scale. The minimum laser spot size achievable when utilizing a visible or near-IR laser is ca. 1 µm, as dictated by the diffraction limit. It is essential to perform some other characterization technique (for instance, scanning electron microscopy combined with atomic resolution atomic force microscopy) to ensure that only a single SWCNT (as opposed to several separate SWCNTs, or a small bundle of SWCNTs) is in the area illuminated by the incident laser.

To assign the (n,m), we begin by noting that the energy gap E_{ii} between one of the pairs of van Hove singularities in the electronic density of states in the SWCNT must lie close to the laser excitation energy E_l for a spectrum to be observed. This is done by comparing E_l to a plot of E_{ii} values for different SWCNT as a function of diameter, known as a Kataura plot ([8] of Chapter 1 and [3] of Chapter 5). It is now known (as discussed in Chapter 5) that the optical resonances in SWCNTs are excitonic in nature. As a result, there is a small reduction in all values of E_{ii} compared to the actual gap between the valence and conduction bands. Since the exciton field lines extend outside of the SWCNT, the value of this small reduction is dependent on the tube environment ([13,15] of Chapter 5). However (as we shall see), many values of E_{ii} for different tubes lie in close proximity (especially for larger SWCNTs). So it is essential to use a Kataura plot which accounts for this environmental effect appropriately for the SWCNT sample under study. For instance, one can choose a Kataura plot for isolated SWCNTs grown vertically on a substrate ([15] of Chapter 5), for isolated SWCNTs deposited on a SiO_2 substrate [21], for solvent-coated SWCNTs in solution [24], or for isolated SWCNTs deposited with sections crossing apertures in a SiO_2 substrate ([17] of Chapter 5).

The range of values that E_l can take relative to E_{ii} for a Raman peak to be observed is known as the resonance window. For the RBM and G peak the resonance window is approximately equal to the energy width of the van Hove singularity plus the energy of the phonon being created. This is a result of the single-resonant nature of the Raman scattering process. During the process the system sits in a real energy level (with energy E_{ii}) and a virtual energy level. The virtual energy level can come before or after the creation of the phonon (Figure 7.18). Hence E_l can be slightly higher than E_{ii} and still produce a resonant Raman scattering process.

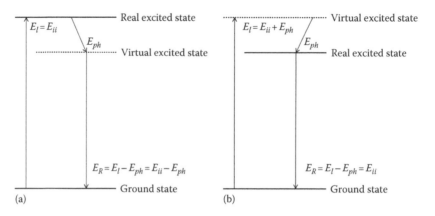

FIGURE 7.18

Schematic energy level diagram illustrating single-resonant Raman scattering transitions responsible for the RBM and G peak in SWCNT Raman spectra. (a) Real excited state (corresponding to E_{ii}) is occupied by system prior to creation of the phonon (energy E_{ph}) and virtual excited state afterwards. (b) Virtual excited state is occupied by system prior to creation of the phonon and real excited state (corresponding to E_{ii}) afterwards. In the latter case, the incident laser photon energy E_l must be slightly larger than E_{ii}.

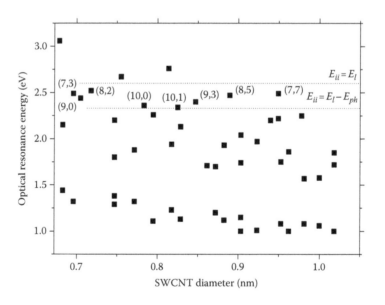

FIGURE 7.19

A Kataura plot with a resonance window for excitation of the Raman G peak with 532 nm excitation superimposed (between the dotted lines). Observation of an SWCNT Raman spectrum under these conditions proves that the SWCNT producing the spectrum lies within the shaded resonance window. (From Liu, K. et al., *Nat. Nanotechnol.*, 7, 325, 2012 [Chapter 5].)

For a given laser excitation energy E_l and a given phonon energy E_{ph}, the energy of the Raman scattered photon E_R is also fixed ($E_R = E_l - E_{ph}$ for Stokes Raman scattering), but E_{ii} can vary within the following resonance window:

$$E_l - E_{ph} \leq E_{ii} \leq E_l \qquad (7.4)$$

The total resonance window is typically ca. 260 meV for the G peak [24] and ca. 100 meV for the RBM [21]. Therefore, it is possible for the G peak of an SWCNT to be within the resonance window whilst the RBM is not.

Figure 7.19 demonstrates a resonance window superimposed onto a Kataura plot. If we observe a Raman spectrum from an SWCNT sample for which this Kataura plot is valid, with this laser excitation, then the tube(s) producing the spectrum must lie within the resonance window shown. Measurement of the RBM frequency, and/or various properties of the G peak, can then be performed to whittle down the possible (n,m) values to a small range, often consisting of just one possible (n,m) assignment.

Whilst this methodology can usually uniquely identify the (n,m) assignment of some individual SWCNTs in a sample, it is important to note that any given laser wavelength can only excite resonant Raman scattering in a small proportion of SWCNTs. Therefore, to draw statistically valid conclusions about the properties of an SWCNT sample containing tubes with a variety of (n,m) assignments, it is essential to perform resonant Raman scattering with a variety of laser wavelengths to excite all tubes expected to be present in the sample.

References

1. F Tuinstra and JL Koenig, *J. Chem. Phys.* **53**, 1126 (1970).

2. H Hiura, TW Ebbesen, K Tanigaki and H Takahashi, *Chem. Phys. Lett.* **202**, 509 (1993).

3. AC Ferrari et al., *Phys. Rev. Lett.* **97**, 187401 (2006).

4. AC Ferrari and DM Basko, *Nat. Nanotechnol.* **8**, 235 (2013).

5. A Jorio, MA Pimenta, AG Souza Filho, R Saito, G Dresselhaus and MS Dresselhaus, *New J. Phys.* **5**, 139.1 (2003).

6. TMG Mohiuddin et al., *Phys. Rev. B* **79**, 205433 (2009).

7. AC Ferrari and J Robertson, *Phys. Rev. B* **61**, 14095 (2000).

8. T An Nguyen, J-U Lee, D Yoon and H Cheong, *Sci. Rep.* **4**, 4630 (2014).

9. Y Wang et al., *J. Phys. Chem C* **112**, 10637 (2008).

10. PH Tan et al., *Nat. Mater.* **11**, 294 (2012).

11. RJ Nemanich, G Lucovsky and SA Solin, *Proceedings of the International Conference on Lattice Dynamics*, pp. 619–621 (Flammarion, Paris, France 1975).

12. M Hanfland, H Beister and K Syassen, *Phys. Rev. B* **39**, 12598 (1989).

13. P Lespade, A Marchand, M Couzi and F Cruege, *Carbon* **22**, 375 (1984).

14. N Wada and S Solin, *Physica B* **105**, 353 (1981).

15. R Robertson and JJ Fox, *Nature* **126**, 279 (1930).

16. K Sato et al., *Chem. Phys. Lett.* **427**, 117 (2006).

17. MM Lucchese et al., *Carbon* **48**, 1592 (2010).

18. J Maultzsch, S Reich and C Thomsen, *Phys. Rev. B* **64**, 121470(R) (2001).

19. MJ Longhurst and N Quirke, *J. Chem. Phys.* **124**, 234708 (2006).

20. PT Araujo et al., *Phys. Rev. B* **77**, 241403(R) (2008).

21. D Zhang et al., *Nanoscale* **7**, 10719 (2015).

22. T Michel et al., *Adv. Nat. Sci.: Nanosci. Nanotechnol.* **1**, 045007 (2010); T Michel et al., *Phys. Rev. B* **80**, 245416 (2009).

23. EH Hároz, JG Duque, WD Rice, CG Densmore, J Kono and SK Doorn, *Phys. Rev. B* **84**, 121403(R) (2011).

24. H Telg et al., *ACS Nano* **6**, 904 (2012).

25. F Herziger, A Vierck, J Laudenbach and J Maultzsch, *Phys. Rev. Lett.* **92**, 235409 (2015).

26. H Wang et al., *ACS Nano* **7**, 614 (2013).

27. JR Sanchez-Valencia et al., *Nature* **512**, 61 (2014).

28. PT Araujo, PBC Pesce, MS Dresselhaus, K Sato, R Saito and A Jorio, *Physica E* **42**, 1251 (2010).

8

Diffraction and Microscopy Experiments on Graphene and Carbon Nanotubes

Electron diffraction has played a large role in the characterization of graphene. It has allowed the unambiguous measurement of the thickness of mono-layer and bi-layer graphene; their electron diffraction patterns are qualitatively different [1]. This has been used to corroborate the fast and non-destructive identification of mono-layer and bi-layer graphene samples using Raman spectroscopy ([3] of Chapter 7). Diffraction has also provided clear experimental evidence of the fact that all real graphene samples are covered, on a microscopic scale, in ripples. This property is of fundamental interest as it is a consequence of the fact that a completely flat two-dimensional sheet of graphene is thermodynamically unstable. We shall prove this in Chapter 10. Observations using electron diffraction have been complimented by studies of graphene using atomic resolution transmission electron microscopy (TEM) [1] and atomic resolution scanning tunnelling microscopy (STM) [2].

In the case of carbon nanotubes, atomic resolution TEM and STM have also played a key role in verifying our understanding. The structure of single-walled, double-walled and multi-walled carbon nanotubes has been directly verified using these methods, and the dependence of electronic dispersion relation on diameter and (n,m) indices that we predicted in Chapter 5 has been directly verified ([5] of Chapter 5).

In this chapter, we will learn a rigorous mathematical method for the treatment for relating the observed diffraction data to the crystal structure of the sample (the Laue treatment) and understand how electron diffraction experiments on graphene have achieved the ends detailed earlier. We will focus also on the information that can be obtained from TEM and STM studies of graphene and nanotubes, and conclude with a discussion of the more limited role taken by X-ray diffraction in this field.

8.1 Laue treatment of diffraction

Readers will be familiar with the Bragg treatment of diffraction, in which we consider the scattering of the incoming beam by successive planes of atoms in the crystalline sample. The angle θ between the diffracted beam and the plane of atoms from which diffraction is taking place is related to the spacing between successive planes by the formula $n\lambda = 2d \sin \theta$, where n is an integer. However, to access the full potential of diffraction techniques to characterize materials, it is essential to use a more rigorous mathematical method: the Laue method.

In the Laue method (see References [2,5] of Chapter 1), we calculate the amplitude of the diffracted beam by summing the amplitude of diffraction from all atoms in the crystal. The symmetry of the crystal makes the problem mathematically tractable. Whilst the initial mathematics is slightly more complex than that for the Bragg method predicting diffraction patterns is generally easier, even for simple structures. Whilst the Bragg method requires the ability to visualize the different "planes" of atoms in the material, with the Laue method, we can simply enter the values of the lattice and basis vectors and turn the mathematical handle and the answer emerges. In addition, using the Laue method we can account for the characteristics of the scattering from the individual atoms in the crystal using a parameter called the structure factor. This allows us to easily predict the relative intensity of the diffracted beams.

8.1.1   Atomic form factor

We must begin by understanding the scattering of radiation* from individual atoms. In a typical atom, the nucleus is $\sim 10^{-15}$ m in diameter and the electron cloud surrounding it is $\sim 10^{-10}$ m in diameter. Since interatomic bonds are formed from the electron clouds in adjacent clouds overlapping, interatomic distances are also $\sim 10^{-10}$ m. We perform a diffraction experiment primarily to measure the distances between the atoms, and hence it is necessary to use a probe with a wavelength (or de Broglie wavelength) similar to the diameter of the atomic electron cloud, $\sim 10^{-10}$ m. In this chapter, we are primarily interested in electron and X-ray diffraction, since these are the probes most often used to characterize graphene and nanotube-related materials. X-rays scatter almost entirely from the electron cloud surrounding the nucleus, whilst scattering of electrons is partially from the electron cloud and partially from the nucleus. In both cases, we assume elastic scattering, and in both cases the scattering from the electron cloud results in a scattered beam with much larger amplitude for smaller scattering angle.

This is a result (Figure 8.1) of the fact that the electron/X-ray beam is scattered from an object of comparable diameter to its de Broglie wavelength. When the scattering angle $2\theta^\dagger$ is small,

---

* In this chapter, we consider the scattering of electrons accelerated so that their de Broglie wavelength is the same order of magnitude as the interatomic spacing in solids and of X-rays.
† We define the scattering angle as 2θ for consistency with the Bragg formula $2d \sin \theta = n\lambda$.

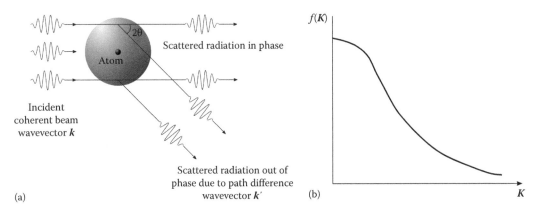

Incident
coherent beam
wavevector **k**

Scattered radiation in phase

Scattered radiation out of
phase due to path difference
wavevector **k**′

(a)

(b)

FIGURE 8.1

(a) Elastic scattering of coherent radiation from a single atom. Radiation scattered through a small angle from different parts of the atom remains in phase so interferes constructively. Radiation scattered through a large angle (large scattering vector **K** = **k**′−**k**) from different parts of the atom interferes destructively due to the large path difference. (b) Sketch of the atomic form factor $f(K)$ as a function of scattering vector **K**, encapsulating this information.

scattered beams from different parts of the atom will be largely in phase, but when the scattering angle is large the optical path difference between the beams scattered from different parts of the atom will be comparable to the (de Broglie) wavelength – so destructive interference will occur and the amplitude of the scattered beam will be lower.

This property of the scattering is encapsulated in a parameter called the atomic form factor f (Figure 8.1). The atomic form factor f describes how the scattering amplitude decreases as the scattering angle increases. In a covalently bonded material such as graphene or diamond, it also depends to some extent on the spatial orientation of the interatomic bonds; this determines where the electrons from which the beam is scattering are likely to be found. In an ionically bonded material the negatively charged ions scatter the beam more strongly as a result of their extra electron(s), and in all cases there is a strong dependence on atomic number. In addition, f is different for X-ray and electron scattering and varies depending on the (de Broglie) wavelength λ of the incoming beam.

Mathematically, the atomic form factor f is approximately proportional to $(\lambda/\sin\theta)$ [3]. We usually write it instead as a function of the scattering vector **K**. This is the change in wavenumber of the electrons/X-rays upon scattering:

$$K = k' - k$$

Assuming elastic scattering ($k' = k = (2\pi/\lambda)$), we can directly relate K to the scattering angle 2θ:

$$K = 2k\sin\theta$$

Assuming for simplicity that f depends only on the scalar values of θ and K, we can write

$$f \propto \frac{4\pi}{K} \tag{8.1}$$

Therefore, whilst we write the atomic form factor as a single parameter f for the sake of brevity, we must remember that in actual fact its value varies significantly depending on a large number of different factors.

8.1.2   Diffraction from a crystal

We continue our study of diffraction by outlining some aspects of the experimental geometry in a diffraction experiment (Figure 8.2). We utilize monochromatic radiation with wavevector k incident on a sample at the origin. The atoms in the sample are at positions r_n relative to the origin. We consider a detector, where the distance from the sample to the detector is large compared to the sample size. Typically (in electron or X-ray diffraction) a sample is illuminated by a beam 1–10 μm in diameter, and the detector is placed many centimetres away from the sample.

To start with, we consider the amplitude of the beam diffracted from the nth atom at a position on the detector defined by the vector R. The scattered beam from the nth

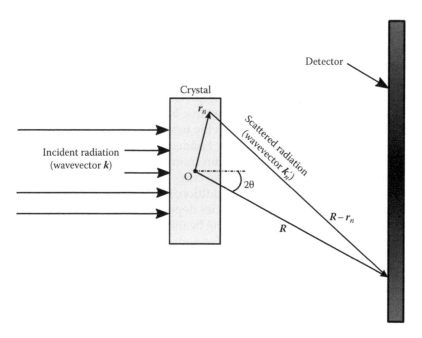

FIGURE 8.2

Experimental geometry for a diffraction experiment involving elastic scattering of a monochromatic coherent beam. We consider the scattering from individual atoms at locations r_n, and the diffracted amplitude at location R on the detector.

atom is hence travelling to this position in the direction $(R - r_n)$ with a wavevector k'_n.

We can write down an expression for the amplitude of the diffracted beam at R from the atom at r_n:

$$E_n = E_0 e^{i(k \cdot r_n - \omega t)} \times f_n \times \frac{e^{ik|R - r_n|}}{|R - r_n|} \tag{8.2}$$

The first term in Equation 8.2 represents the amplitude of the incident beam, and the second is the atomic form factor for the nth atom. The third represents the phase change (numerator) and amplitude decrease (denominator) upon scattering. Recalling that $k = k'_n$ and that $(R - r_n)$ and k'_n point in the same direction, we can manipulate the phase factor as follows:

$$k|R - r_n| = k'_n \cdot (R - r_n) = k'_n \cdot R - k'_n \cdot r_n \approx kR - k'_n \cdot r_n$$

We are now in a position to introduce the scattering vector K for the diffracted beam from the nth atom:

$$K = k'_n - k \tag{8.3}$$

and can manipulate Equation 8.2 to obtain

$$E_n = E_0 e^{i(k \cdot r_n - \omega t)} \times f_n \times \frac{e^{ikR} e^{-ik'_n \cdot r_n}}{|R - R_n|}$$

$$E_n = \frac{E_0 e^{i(kR - \omega t)}}{R} \times f_n \times e^{-iK \cdot r_n} \tag{8.4}$$

where we have also used the approximation $|R - r_n| \cong R$, since the detector is situated a long distance away from a small sample. To obtain the amplitude of the diffracted beam, we sum the amplitude of the scattered beam given by Equation 8.4 for all atoms in the crystal. The first term, however, is not of interest as it is the same for all atoms in the crystal. We hence obtain an expression for the total diffracted amplitude:

$$E = \sum_n f_n e^{-iK \cdot r_n} \tag{8.5}$$

To proceed, we split this summation into two summations: one over the atoms that form the unit cell and the other over the unit cells that form the crystal. So r_n is defined as follows: $r_n = r_l + r_b$. r_l give the location of the lattice points (i.e. they are the lattice vectors) and r_b give the location of the atoms in the unit cell, relative to the lattice point (i.e. they are the basis vectors). We can therefore obtain

$$E(K) = \sum_l e^{-iK \cdot r_l} \times \sum_b f_b e^{-iK \cdot r_b} \tag{8.6}$$

We have written $E(K)$ because we are measuring the diffracted amplitude as a function of scattering angle 2θ, that is, as a function of K. The first term in Equation 8.6 is a summation of the diffracted amplitude from all the unit cells in the crystal. This term determines the values of K (or 2θ) for which a diffracted beam can be present. The second term in Equation 8.6 determines the relative amplitude of the different diffracted beams and can cause certain diffracted beams to have amplitude of zero. This term is called the structure factor.

8.1.3   The reciprocal lattice

Before we deal with the structure factor, we need to use the first term to elucidate what special values of the scattering vector K lead to a diffracted beam being present. Let us consider a crystal with three lattice vectors a, b, c. r_l can be written as integer multiples of these lattice vectors: $r_l = p_1 a + p_2 b + p_3 c$. The first summation in Equation 8.6 can hence be divided as follows:

$$\sum_l e^{-iK\cdot r_l} = \sum_{p_1} e^{-ip_1 K \cdot a} \sum_{p_2} e^{-ip_2 K \cdot b} \sum_{p_3} e^{-ip_3 K \cdot c} \qquad (8.7)$$

By inspection of Equation 8.7, we can see that there can be an intense diffracted beam when all the exponential terms have a value of 1. To achieve this $K \cdot a$, $K \cdot b$ and $K \cdot c$ must all be integer multiples of 2π. These conditions are the Laue conditions for diffraction and can be stated as follows:

$$K \cdot a = 2\pi h \qquad (8.8)$$

$$K \cdot b = 2\pi k$$

$$K \cdot c = 2\pi l$$

where h, k, l are integers. Any diffracted beam can therefore be uniquely labelled using the integers h, k, l. These are referred to in the literature as the Miller indices of that diffraction peak. Furthermore, the allowed values of K that lead to a diffracted beam form a lattice in reciprocal space. This is the reciprocal lattice, the same lattice of points in reciprocal space that determines the periodicity of the electron and phonon dispersion relations that you have studied in previous chapters. We can relate the scattering vector to a reciprocal lattice point G_{hkl} and the reciprocal lattice vectors a^*, b^*, c^* as follows:

$$K = G_{hkl} = ha^* + kb^* + lc^* \qquad (8.9)$$

Readers may wish to check for themselves that the following expressions for the reciprocal lattice vectors always result in reciprocal lattice vectors that satisfy the Laue conditions Equation 8.8:

$$a^* \cdot a = b^* \cdot b = c^* \cdot c = 2\pi \qquad (8.10)$$

$$b^* \cdot a = c^* \cdot a = a^* \cdot b = c^* \cdot b = a^* \cdot c = b^* \cdot c = 0 \qquad (8.11)$$

Furthermore, once the real space lattice vectors have been defined, there is only one set of reciprocal lattice vectors that will satisfy Equations 8.10 and 8.11. This is true whether we are considering a two-dimensional material (with two real space lattice vectors and two reciprocal space lattice vectors) or a three-dimensional material (with three real space lattice vectors and three reciprocal space lattice vectors). We applied these conditions (8.10) and (8.11) to calculate the graphene reciprocal space lattice vectors from the real space lattice vectors in Chapter 1.

But if we are dealing with a three-dimensional material, we can achieve exactly the same end mathematically by defining the reciprocal lattice vectors in terms of the real space lattice vectors as follows:

$$a^* = \frac{2\pi(b \times c)}{a \cdot (b \times c)}; \quad b^* = \frac{2\pi(c \times a)}{a \cdot (b \times c)}; \quad c^* = \frac{2\pi(a \times b)}{a \cdot (b \times c)} \qquad (8.12)$$

Matters are simplified by the fact that $a \cdot (b \times c)$ is the volume of the real space unit cell. Hence (assuming the real space lattice extends infinitely), we can envisage the reciprocal lattice as an infinite lattice in reciprocal space of all the scattering vectors resulting in a diffraction peak which could, in principle, be observed.

Using our methodology so far, we know how to measure the reciprocal lattice vectors and their relation to the real space lattice vectors. But this would tell us nothing about what lies at each lattice point (the contents of the unit cell). To learn about this, we need to measure the relative intensities of each diffraction peak, which is determined by what lies in each unit cell via the structure factor.

8.1.4   The structure factor

Now that we have calculated the values of the scattering vector K for which there will be a diffracted beam (Equation 8.9), we are in a position to evaluate the structure factor. Physically, the structure factor S accounts for the fact that depending on the value of K, the beams diffracted from the various atoms in the unit cell may interfere constructively or destructively. It also, by incorporating the atomic form factor, accounts for the specific nature of the scattering from each individual atom as discussed earlier. To recap on the relevant part of Equation 8.6, the structure factor is given by

$$S = \sum_b f_b e^{-iK \cdot r_b} \qquad (8.13)$$

where the vectors r_b are the basis vectors: the locations of the atoms in the unit cell* relative to the lattice point. Usually in the literature these vectors are written as fractions of the lattice vectors. Inspection of Equation 8.13 reveals why this is a good idea; it allows us to easily evaluate the exponents using the Laue conditions (Equation 8.8). Thus, to predict the relative intensity of different diffracted beams, we simply use the (hkl) values and the reciprocal lattice vectors to define K, then evaluate S for each (hkl).

Since the structure factor depends on the contents of the unit cell, it is different for mono-layer and bi-layer graphene. We will evaluate it later for both these cases.

8.1.5 Probing reciprocal space through diffraction

We noted in Chapter 1 that the reciprocal space lattice is the Fourier transform of the real space lattice [4]. Readers are referred to Appendix C for a detailed demonstration of this fact. When we characterize a diffraction spot (calculate the scattering vector K that has caused the spot, measure the amplitude of the spot), we are characterizing a point on the reciprocal lattice.

In principle, we could therefore unambiguously determine the real space lattice of a crystal by conducting a diffraction experiment in which we measured the reciprocal lattice over all reciprocal space, then performed an inverse Fourier transform on the data to directly determine the real space lattice.

Unfortunately, in a real diffraction experiment we do not probe all of reciprocal space. The experiment is often a battle against the various experimental constraints to probe as much reciprocal space as possible. We collect diffraction pattern(s) that cover a portion of reciprocal space, then we take an educated guess at what the (real space) structure might be. We use the methodology outlined so far to predict the diffraction pattern in the portion of reciprocal space that we measured during the experiment. Then, we adjust the parameters in our model (such as the real space lattice constants) to get a result more closely matching the actual diffraction data. This is usually done using a least-squares fitting process such as Rietveld refinement [5]. Usually, there are very many parameters to adjust but only one or two can be automatically adjusted simultaneously otherwise the program will be unable to find a physically meaningful minimum in the least-squares fit. So it is necessary to painstakingly adjust some parameters manually and allow automatic adjustment (refinement) of other parameters in successive runs of the least-squares fitting routine. This process is called structural refinement and many scientists spend their entire careers doing it.

---

* This does not have to be the primitive unit cell. We can choose any unit cell that we like, as long as we choose appropriate lattice vectors to go with the unit cell and basis vectors r_b.

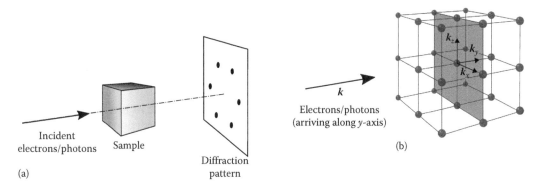

FIGURE 8.3

An electron diffraction experiment represented in real space (a) and reciprocal space (b). In real space, we collect a single two-dimensional diffraction pattern. In reciprocal space, this corresponds to sampling the plane through the reciprocal lattice lying perpendicular to the incident beam.

So how do we probe as much reciprocal space as possible in a diffraction experiment? Let us first consider what happens when we collect a single diffraction image of a single crystal sample (Figure 8.3).

Suppose we fire a monochromatic beam of electrons/X-ray photons with wavevector k at a single crystal sample. All possible values of G_{hkl} form a lattice in reciprocal space. All of these lattice points can give rise to a diffracted beam if the geometry of the experiment and value of the atomic form factor permit it.

When we detect the electrons/photons downstream, we do not measure the energy of the electrons/photons (in fact, we assume that scattering is elastic). We detect the angle through which they are scattered. In the setup shown in figure 8.3, we only measure values of G_{hkl} in the xz plane of reciprocal space.*

We can also observe that each diffraction pattern collected is a two-dimensional dataset so we cannot expect to construct a three-dimensional lattice from a single pattern.

If we want to make measurements outside the xz plane, we need to rotate the sample relative to the beam of incoming electrons/photons. Generally, it is more convenient to rotate the sample rather than the beam but experimental constraints usually prevent us from rotating the sample as much as we would like to. For instance, if we were to perform electron diffraction on graphene supported by a substrate with an aperture then we might find that the edges of the aperture started blocking the incoming beam when we rotated the sample by a large angle.

---

* This is a simplification which works well for an electron diffraction experiment such as those discussed in this chapter. In actual fact, the plane described is a small part of a sphere in reciprocal space called the Ewald sphere. The interested reader is referred to texts treating the subject in more detail (e.g. [5] of Chapter 1).

In addition, when we perform the single diffraction experiment shown in Figure 8.3, we do not sample the entire xz plane because the detector only occupies a small proportion of the cross-sectional area downstream of the sample, and/or because the atomic form factor takes an increasingly low value for large K.

These two constraints ensure that in any real diffraction experiment we only sample a small portion of the reciprocal lattice.

8.2 Reciprocal lattice and structure factor of graphene

The two-dimensional real space lattice vectors of graphene were given in Chapter 1 and are determined by the hexagonal symmetry of the lattice. To recap, these lattice vectors are

$$\boldsymbol{a}_1 = \begin{pmatrix} \dfrac{\sqrt{3}a}{2} \\ \dfrac{a}{2} \end{pmatrix}; \quad \boldsymbol{a}_2 = \begin{pmatrix} \dfrac{\sqrt{3}a}{2} \\ -\dfrac{a}{2} \end{pmatrix} \quad \text{where } a = \sqrt{3}a_0 \qquad (8.14)$$

We expect therefore two reciprocal lattice vectors $\boldsymbol{b}_1$ and $\boldsymbol{b}_2$. For $\boldsymbol{b}_1$, the conditions $\boldsymbol{a}_1 \cdot \boldsymbol{b}_1 = 2\pi$ and $\boldsymbol{a}_2 \cdot \boldsymbol{b}_1 = 0$ provide two simultaneous equations from which the two components of $\boldsymbol{b}_1$ can be uniquely determined. Performing this process for $\boldsymbol{b}_1$ and $\boldsymbol{b}_2$ leads to the following expressions:

$$\boldsymbol{b}_1 = \begin{pmatrix} \dfrac{2\pi}{\sqrt{3}a} \\ \dfrac{2\pi}{a} \end{pmatrix}; \quad \boldsymbol{b}_2 = \begin{pmatrix} \dfrac{2\pi}{\sqrt{3}a} \\ -\dfrac{2\pi}{a} \end{pmatrix} \qquad (8.15)$$

We therefore expect to observe diffraction peaks from a graphene crystal at the following scattering vectors:

$$\boldsymbol{K} = h\boldsymbol{b}_1 + k\boldsymbol{b}_2 \qquad (8.16)$$

In the case of a two-dimensional material such as graphene, the third Miller index l is redundant. By inspection of Equations 8.15 and 8.16, we can identify the Miller indices of an electron diffraction image of mono-layer graphene (Figure 8.4). The resemblance to the reciprocal lattice that we have calculated is clear.

8.2.1 Structure factor of mono-layer graphene

The evaluation of the structure factor allows us to unambiguously identify an electron diffraction pattern as emanating from mono-layer graphene, as distinct from bi-layer or few-layer graphene or (for that matter) graphite.

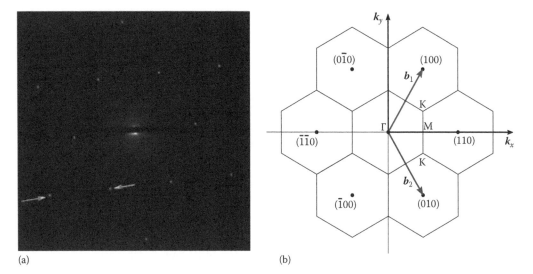

(a)   (b)

FIGURE 8.4
(a) Normal incidence electron diffraction pattern of graphene. (Reprinted with permission from Ferrari, A.C. et al., *Phys. Rev. Lett.*, 97, 187401 [Chapter 7]. Copyright 2006, American Physical Society.) (b) Schematic diagram showing the graphene reciprocal lattice with points labelled according to the Miller indices (*hkl*).

Let us make the comparison between mono-layer and bi-layer graphene. The lattice vectors are the same for both mono-layer and bi-layer graphene, hence we expect to observe the same diffraction spots. However, having chosen the same lattice vectors, we have no choice but to define different unit cells for mono-layer and bi-layer graphene. The unit cell of mono-layer graphene consists of two atoms whilst the unit cell of bi-layer graphene consists of four atoms, two in each layer. Hence, the structure factor is different and the relative intensity of the diffraction spots is different.

For mathematical convenience, we will assemble the two-atom unit cell of mono-layer graphene from the two atoms joined by a bond along the x-axis with one at the origin. The basis vectors are therefore

$$r_1 = \begin{pmatrix} 0 \\ 0 \end{pmatrix} \tag{8.17}$$

$$r_2 = \begin{pmatrix} \dfrac{a}{\sqrt{3}} \\ 0 \end{pmatrix} = \dfrac{a_1 + a_2}{3}$$

The expression for the structure factor (Equation 8.13) therefore yields

$$S_1 = f\left[e^{-iK\cdot r_1} + e^{-iK\cdot r_2} \right] = f\left[1 + e^{-\frac{2\pi i}{3}(h+k)} \right] \tag{8.18}$$

where, to obtain this expression, we have assumed that the atomic form factor f is the same for both atoms. We have used Equation 8.16 and the Laue conditions for diffraction (Equation 8.8) to evaluate the product $\mathbf{K} \cdot \mathbf{r}_2$. From inspection of Equation 8.18, we can see that the value of the structure factor will vary according to the values of h and k, but there are no values of h and k that will result in $S_1 = 0$. Hence, there are no "missing" diffraction spots from mono-layer graphene.

To compare the intensities of the diffraction peaks (as opposed to the amplitudes) we need to use $|S_1|^2$:

$$|S_1|^2 = 2f^2 \left[1 + \cos\left[\frac{2\pi}{3}(h+k) \right] \right] \tag{8.19}$$

8.2.2   Structure factor of bi-layer graphene

In bi-layer graphene there are four atoms in the unit cell. Their locations depend on the nature of the stacking, but we shall use Bernal (or ABAB) stacking, the most common and thermodynamically stable form. In this case, the four atoms composing the unit cell are at the following locations:

$$\mathbf{r}_1 = \begin{pmatrix} 0 \\ 0 \\ 0 \end{pmatrix} \tag{8.20}$$

$$\mathbf{r}_2 = \begin{pmatrix} \dfrac{a}{\sqrt{3}} \\ 0 \\ 0 \end{pmatrix} = \frac{\mathbf{a}_1 + \mathbf{a}_2}{3}$$

$$\mathbf{r}_3 = \begin{pmatrix} 0 \\ 0 \\ c \end{pmatrix}$$

$$\mathbf{r}_4 = \begin{pmatrix} -\dfrac{a}{\sqrt{3}} \\ 0 \\ c \end{pmatrix} = c\hat{\mathbf{z}} - \frac{\mathbf{a}_1 + \mathbf{a}_2}{3}$$

Unfortunately, since there are no lattice vectors in the $\hat{z}$ direction, it is not possible to express $\mathbf{r}_3$ and $\mathbf{r}_4$ completely in terms of the lattice vectors. But since $\mathbf{b}_1$, $\mathbf{b}_2$ and hence $\mathbf{K}$ are in the xy plane, the $\hat{z}$ component of $\mathbf{r}_3$ and $\mathbf{r}_4$ will disappear when we take the scalar product with $\mathbf{K}$. We therefore obtain

$$S_2 = f\left[e^{-iK \cdot r_1} + e^{-iK \cdot r_2} + e^{-iK \cdot r_3} + e^{-iK \cdot r_4}\right]$$

$$S_2 = 2f\left[1 + \cos\left[\frac{2\pi}{3}(h+k)\right]\right]$$

$$|S_2|^2 = 4f^2\left[1 + \cos\left[\frac{2\pi}{3}(h+k)\right]\right]^2 \qquad (8.21)$$

8.2.3   Identification of graphene sample thickness via electron diffraction

We are now ready to identify the thickness of different graphene samples using electron diffraction. Figure 8.5 shows two-dimensional electron diffraction patterns of mono-layer and bi-layer graphene samples, and a plot of the integrated intensity of selected peaks.

Using our expressions for the structure factors of mono-layer and bi-layer graphene (Equations 8.19 and 8.21) and for the variation of atomic form factor as a function of scattering vector (Equation 8.1), we can show (see problems) that the

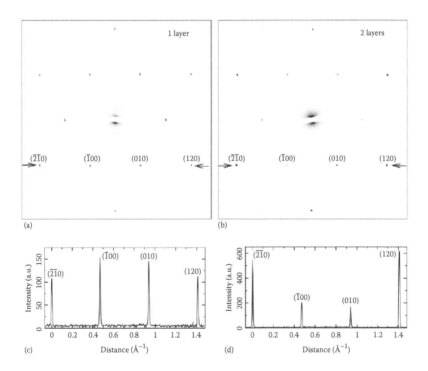

FIGURE 8.5
Two-dimensional electron diffraction images of mono-layer (a) and bi-layer (b) graphene samples. (c and d) Integrated intensities of diffraction spots along the horizontal lines marked between the arrows in the two-dimensional images. (Reprinted from *Solid State Commun.*, 143, Meyer, J.C. et al., 101, Copyright 2007, with permission from Elsevier.)

intensity ratios of the (010) and (120) diffraction peaks should be different for the case of mono-layer and bi-layer graphene:

$$\text{Mono-layer}: \frac{I(010)}{I(120)} = \frac{3}{4} \qquad (8.22)$$

$$\text{Bi-layer}: \frac{I(010)}{I(120)} = \frac{3}{16} \qquad (8.23)$$

Here, we have successfully explained the drastic increase in intensity of the (120) peak relative to the (010) in the bi-layer case (shown in Figure 8.5). Note, however, that Equations 8.22 and 8.23 are only qualitatively correct. In both mono-layer and bi-layer graphene, the actual data in Figure 8.5 show the (120) peak less intense than it should be according to Equations 8.22 and 8.23. Why is this? It is because we have used an approximate formula (8.1) for the atomic form factor.

Whilst comparison of the intensity of the various $(hk0)$ reflections provides substantive evidence to distinguish between mono-layer and bi-layer graphene, we can learn more by looking beyond the xy plane in reciprocal space; In this case, we have to make a modification to our treatment of diffraction earlier to account for the two-dimensional nature of graphene. Rewriting Equations 8.6 and 8.7, the expressions from which the Laue conditions were obtained, for the specific case of graphene, we obtain the expression for the amplitude of the diffracted beam as a function of K:

$$E(K) = \sum_{p_1} e^{-ip_1 K \cdot a_1} \sum_{p_2} e^{-ip_2 K \cdot a_2} \times \sum_b f e^{-iK \cdot r_b} \qquad (8.24)$$

To obtain a large amplitude K must be a vector resulting in all the exponential terms in the first two summations being equal to unity, that is, $K \cdot a_1 = 2\pi h$ and $K \cdot a_2 = 2\pi k$. However, since a_1 and a_2 both lie in the xy plane K can also have a z-component which takes any value without affecting the values of $K \cdot a_1$ and $K \cdot a_2$. So we can write the vectors K satisfying the Laue conditions for diffraction as follows:

$$K = hb_1 + kb_2 + k_z \hat{z} \qquad (8.25)$$

where k_z is a continuous variable that can take any value. But to obtain a diffracted beam the structure factor (the last summation in Equation 8.24) must also take a non-zero value. For the case of mono-layer graphene, this is relatively simple to evaluate. Both the basis vectors r_p (given by Equation 8.17) lie in the xy plane so the z-component of K will not affect the value of the structure factor; until k_z (and hence K) become so large that the value of the atomic form factor f drops as a result.

However, for the case of a bi-layer the value of the structure factor oscillates as a function of k_z, because two of the basis vectors (Equations 8.20) have components in the $\hat{z}$ direction. The bi-layer structure factor S_2 can be evaluated as follows:

$$S_2 = \sum_b f e^{-iK \cdot r_b} = f \left[e^{-iK \cdot r_1} + e^{-iK \cdot r_2} + e^{-iK \cdot r_3} + e^{-iK \cdot r_4} \right]$$

where (in contrast to our treatment leading to Equation 8.21) we must consider the z-components in the scalar products in the last two exponents. We hence obtain

$$S_2 = f \left[1 + e^{-\frac{2i\pi}{3}(h+k)} + e^{-ik_zc} + e^{-ik_zc} e^{\frac{2i\pi}{3}(h+k)} \right] \qquad (8.26)$$

The value of the bi-layer structure factor S_2 therefore oscillates as k_z varies, even dropping to zero.

The results of a more detailed calculation [1] (taking into account, for instance, the expected variation of the atomic form factor f as a function of K) of the values of K for which a diffracted beam will be present are shown in Figure 8.6. Just as in our simple calculation earlier, the detailed calculation predicts an oscillating value of the structure factor as a function of k_z for a bi-layer, but not for a mono-layer.

The diffraction data shown in Figure 8.5 were collected with the incoming beam perpendicular to the graphene lattice, therefore sampling the xy plane in the reciprocal lattice of graphene (horizontal plane in Figure 8.6). To observe the z-component of K, we need to sample as much of the reciprocal lattice as possible beyond the xy plane by rotating the graphene sample relative to the incoming electron beam. The results of

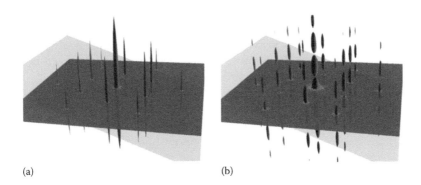

(a)   (b)

FIGURE 8.6
Expected values of K for which peaks will be present in the electron diffraction patterns of mono-layer (a) and bi-layer (b) graphene. The horizontal plane is the xy plane and the slanted plane is an example of another plane that can be sampled, by tilting the graphene sheet relative to the electron beam. The thickness (in the xy plane) of the various allowed values of K represents the intensity of the peak, not the width. (Reprinted from *Solid State Commun.*, 143, Meyer, J.C. et al., 101, Copyright 2007, with permission from Elsevier.)

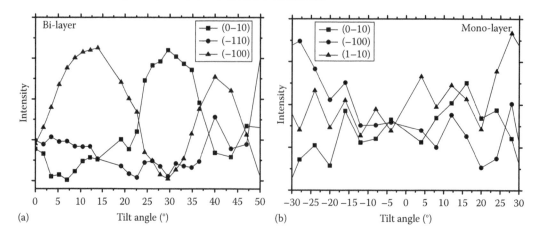

FIGURE 8.7

Intensity of some electron diffraction peaks from graphene as a function of the tilt angle between the graphene sample and the incident electron beam. The intensity of the peaks from the bi-layer sample (a) oscillates, an effect absent from the mono-layer sample (b). (Reprinted from *Solid State Commun.*, 143, Meyer, J.C. et al., 101, Copyright 2007, with permission from Elsevier.)

such an experiment are shown in Figure 8.7. The intensity of the diffraction peaks from the bi-layer sample oscillates; an effect absent from the mono-layer case, as expected.

8.3   Electron diffraction on graphene: Evidence for microscopic rippling of the graphene sheet

We have already discussed in previous chapters the fact that all real graphene sheets are, on a microscopic level, heavily rippled. In Chapter 10, we will discover the thermodynamical reason why this is the case. The ripples always observed in real graphene sheets have been characterized in detail using electron microscopy.

We demonstrated earlier that values of the scattering vector given by $K = hb_1 + kb_2 + k_z\hat{z}$ can produce a diffracted beam. The scattering vector can take only discrete values in the xy plane but can vary continuously in the z direction, normal to the graphene plane. We can therefore plot (Figure 8.8) the allowed values of K leading to a diffraction peak from a single completely planar graphene sheet as straight lines parallel to the z-axis.

However, the z direction is not quite normal to most areas of a real (rippled) graphene sheet. The allowed values of K from a real graphene sheet will therefore form a cone around an axis parallel to the z-axis (Figure 8.8).

We therefore expect that if we perform a diffraction experiment on a graphene sheet normal to the incident electron beam, then gradually tilt the graphene sheet relative to the

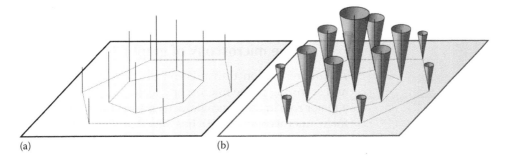

(a)   (b)

FIGURE 8.8
Schematic diagram of allowed values of K leading to a diffraction peak from a hypothetical flat graphene sheet (a) and a real rippled graphene sheet (b).

beam, the diffraction peaks will gradually get broader as we tilt further. The further we tilt the graphene sheet, the wider an area of the cone in Figure 8.8 will be intersected by the plane in reciprocal space that we are sampling. This is exactly what is observed experimentally (Figure 8.9). It is also worth noting that the effect is more prominent for a mono-layer than a bi-layer; evidence that thicker graphene samples have fewer and smaller ripples.

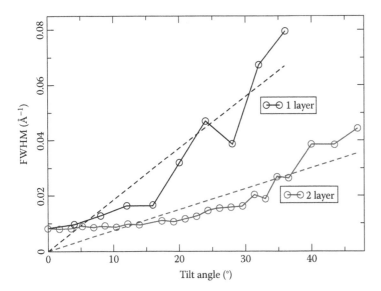

FIGURE 8.9
Full width half maximum of $(0\overline{1}0)$ diffraction peak from mono-layer (left hand panel) and bi-layer (right hand panel) graphene as a function of tilt angle between the graphene sheet and the incident beam. The much larger broadening of the peak from the mono-layer sample is a consequence of the rippling of the graphene sheet. (Reprinted from *Solid State Commun.*, 143, Meyer, J.C. et al., 101, Copyright 2007, with permission from Elsevier.)

8.4 Atomic resolution transmission electron microscopy (TEM) and scanning probe microscopy of graphene

Atomic resolution transmission electron microscopy (TEM) and scanning probe microscopy techniques have also played a major role in the basic characterization of graphene. Atomic force microscopy (AFM) has been used to directly verify sample thickness ever since the initial discovery of graphene ([24] of Chapter 1). Atomic resolution TEM [1], scanning transmission electron microscopy [6] and scanning tunnelling microscopy (STM) [2,7] have been used to directly verify the presence of the hexagonal graphene lattice with lattice constant the same as that of bulk graphite within experimental error. In addition to the use of TEM to study graphene, current research also focusses on the use of graphene as the thinnest possible transparent and conductive support for the imaging of other samples in the TEM [8].

Scanning probe microscopy techniques (STM and AFM) have the advantages over TEM that graphene samples can be imaged on substrates such as SiO_2 [9], SiC [7] and Cu [10], as well as in freestanding form [2]. They can also directly image the surface topography of the sample. Images such as Figure 8.10 can be obtained demonstrating the presence

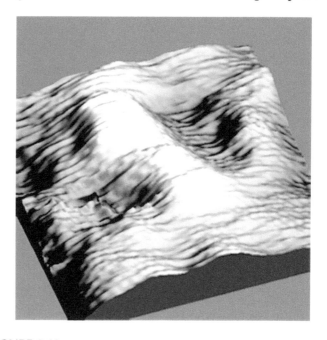

FIGURE 8.10
STM image of 10 nm × 10 nm area of suspended graphene, demonstrating the presence of static ripples in the graphene sheet, at ambient temperature. (From Zan, R. et al., *Nanoscale*, 4, 3065, 2012. Reproduced by permission of the Royal Society of Chemistry.

of (static) ripples with wavelength λ~5 nm in the lattice of suspended graphene, a conclusion supported by scanning transmission electron microscopy [11] and electron diffraction as discussed earlier. Such ripples have also been observed in graphene on a substrate [12].

Scanning probe microscopy techniques also allow the possibility of making other measurements on graphene using the probe, for instance, the intrinsic strength of graphene can be measured by indenting a graphene membrane covering an aperture in a substrate ([34] of Chapter 1), as discussed in Chapter 10.

8.5   Atomic resolution transmission electron microscopy (TEM) and scanning probe microscopy of carbon nanotubes

Electron diffraction has been used for identification of carbon nanotubes [13] but has not been used as widely as in the study of graphene. TEM and scanning probe microscopy techniques have played a slightly different role in the study of carbon nanotubes (both single-walled and multi-walled) due to this, and due to their one-dimensional nature.

In the first instance, both TEM ([14] of Chapter 1) and STM ([5] of Chapter 5 and [14]) provided clear evidence to prove that nanotubes are composed of a hexagonal lattice of carbon atoms with a lattice constant very close to that of graphene and graphite. The number of graphene layers forming the nanotube can be directly counted using TEM. In the case of the STM studies, I-V spectroscopy utilizing the STM tip provides information on the electronic properties of the same tube that is being imaged. Hence the dependence of electronic dispersion relation on diameter and (n,m) indices that we predicted in Chapter 5 can be directly verified ([5] of Chapter 5 and [14,15]).

In addition, TEM has been used extensively to image carbon nanotubes (both single-walled and multi-walled) as part of the bundles in which they are grown. Such studies have yielded a wealth of information about the growth processes, diameter distribution of tubes within the bundle and also stability of individual tubes. Figure 8.11 shows examples of the kind of TEM images that can be collected of carbon nanotubes. The size of the bundles can be measured, nanotubes can be observed growing out of the transition metal nanoparticles used to catalyse the growth process, and the stability of the circular nanotube cross section against collapse to an oval/racecourse shape for different diameter nanotubes can be directly observed. Note that it is often possible to obtain a cross-section-like view of the nanotube bundle. This is a result of the fact that the section of the bent nanotube bundle

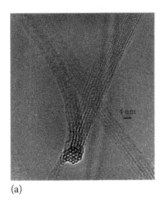
(a)

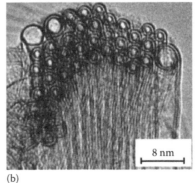
(b)

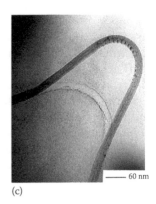
(c)

FIGURE 8.11

A selection of TEM images of carbon nanotubes. (a) Image of a bundle formed from SWCNT. (b) Image of a bundle formed of a mixture of SWCNT, double-walled carbon nanotubes and multi-walled carbon nanotubes. (c) Image of kinks formed in a buckled individual multi-walled carbon nanotube. ([a]: Reprinted by permission from Macmillan Publishers Ltd. *Nature*, Journet, C. et al., 388, 756, Copyright 1997; [b]: Colomer, J.-F., Henrard, L., Van Tendeloo, G., Lucas, A. and Lambin, P., *J. Mater. Chem.*, 14, 603, 2004. Reproduced by permission of the Royal Society of Chemistry; [c]: Reprinted with permission from Lourie, O., Cox, D.M. and Wagner, H.D., *Phys. Rev. Lett.*, 81, 1638. Copyright 1998 by the American Physical Society.)

parallel to the incoming beam will present the largest collision cross section to the beam, not of any actual process of cutting the nanotubes.

Due to their high aspect ratio and atomically thin dimensions, SWCNTs make – arguably – ideal tips for use in scanning probe microscopy. Simple proceduress have been developed to pick up SWCNTs from substrates to manufacture an AFM tip [16,17], and chemically selective imaging is possible through attachment of functional groups to the end of the SWCNT tips [18]. Such tips have been used in AFM studies, particularly of biological molecules [19,20].

8.6 X-ray diffraction studies of graphene and SWCNTs

X-ray diffraction has played a minor role in the study of graphene and carbon nanotubes compared to electron diffraction, electron microscopy and scanning probe microscopy. This is because, generally, good X-ray diffraction data cannot be obtained on graphene and carbon nanotubes for two key reasons. Firstly, the low X-ray scattering cross section of carbon prevents a signal being observed from atomically thin samples. Secondly, the phenomenon of Scherrer broadening (Appendix C) results in the broadening and disappearance of many of the diffraction peaks from atomically thin samples. For a given sample, the phenomenon of Scherrer broadening is more severe in an X-ray diffraction experiment than an electron diffraction experiment due to the longer wavelength of X-rays compared to relativistic electrons. A typical synchrotron X-ray diffraction experiment utilizes X-rays with a wavelength of 4×10^{-11} m [21], whilst a typical electron diffraction experiment utilizes electrons with a de Broglie wavelength of $\sim 10^{-12}$ m. Indeed, it is possible to accelerate electrons to even higher velocity so that their de Broglie wavelength is $\sim 10^{-15}$ m – small enough to diffract just from the atomic nucleus [22].

X-ray diffraction has however provided some useful information about graphene and SWCNT. In Chapter 11, we will learn about the negative thermal expansion coefficient of graphene, a direct consequence of its two-dimensional nature. In graphene, it is very hard – if not impossible – to make an accurate quantitative experimental measurement of the thermal expansion coefficient. But fortunately the interaction between successive atomic layers of graphite is sufficiently weak that even graphite exhibits a negative thermal expansion coefficient (in the in plane direction). This has been measured in an accurate quantitative manner using X-ray diffraction ([3] of Chapter 6 and [23,24]).

X-ray diffraction has also played a role in the study of SWCNTs. Whilst X-ray diffraction peaks from the actual graphene lattice making up the nanotubes are not observed, the nanotubes themselves form a lattice (see Figure 8.11); the lattice with triangular symmetry that is formed from the individual nanotubes packing together into the bundle. A weak and broad X-ray diffraction signal from this lattice can be observed [25].

For both SWCNTs and graphene, X-rays have one key advantage as a probe compared to electrons; their extremely high penetrating power allows their use to probe samples in environments that cannot be accessed with electron beams. For instance, X-ray diffraction experiments on carbon nanotubes [26], few-layer graphene [27], graphite ([12] of Chapter 7) and fullerenes [28] have been performed at ultra-high pressures (over 10,000 atmospheres) inside the diamond anvil high pressure cell. This work forms part of our knowledge and understanding of the mechanical properties of graphene and carbon nanotubes, to which Chapter 10 is devoted.

References

1. JC Meyer et al., *Solid State Commun.* **143**, 101 (2007).

2. R Zan et al., *Nanoscale* **4**, 3065 (2012).

3. DJ Dyson, *X-Ray and Electron Diffraction Studies in Materials Science*, Maney, London, U.K. (2004).

4. HM Rosenberg, *The Solid State*, OUP, Oxford (1975).

5. HM Rietveld, *J. Appl. Cryst.* **2**, 65 (1969).

6. MH Gass, U Bangert, AL Bleloch, P Wang, RR Nair and AK Geim, *Nat. Nanotechnol.* **3**, 676 (2008).

7. P Mallet, F Varchon, C Naud, L Magaud, C Berger and J-Y Veuillen, *Phys. Rev. B* **76**, 041403(R) (2007).

8. JC Meyer, CO Girit, MF Crommie and A Zettl, *Nature* **454**, 319 (2008).

9. E Stolyarova et al., *Proc. Natl. Acad. Sci. USA* **104**, 9209 (2007).

10. L Gao, JR Guest and NP Guisinger, *Nano Lett.* **10**, 3512 (2010).

11. U Bangert, MH Gass, AL Bleloch, RR Nair and AK Geim, *Phys. Stat. Sol.* (a) **206**, 1117 (2009).

12. M Ishigami, JH Chen, WG Cullen, MS Fuhrer and ED Williams, *Nano Lett.* **7**, 1643 (2007).

13. M Gao, JM Zuo, RD Twesten, I Petrov, LA Nagahara and R Zhang, *Appl. Phys. Lett.* **82**, 2703 (2003).

14. TW Odom, J-L Huang, P Kim and CM Lieber, *Nature* **39**, 62 (1998).

15. M Ouyang, J-L Huang, CL Cheung and CM Lieber, *Science* **292**, 702 (2001).

16. JH Hafner, C-L Cheung, TH Oosterkamp and CM Lieber, *J. Phys. Chem. B* **105**, 743 (2001).

17. NR Wilson and JV Macpherson, *Nat. Nanotechnol.* **4**, 483 (2009).

18. T Nishino, T Ito and Y Umezawa, *Anal. Chem.* **74**, 4275 (2002).

19. Y Li et al., *Nat. Mater.* **3**, 38 (2004).

20. N Gadegaard, *Biotech. Biochem.* **81**, 87 (2009).

21. M Hanfland, JE Proctor, CL Guillaume, O Degtyareva and E Gregoryanz. *Phys. Rev. Lett.* **106**, 095503 (2011).

22. WSC Williams, *Nuclear and Particle Physics*, OUP (1991).

23. HO Pierson, *Handbook of Carbon, Graphite, Diamond, and Fullerenes: Properties, Processing, and Applications*, Noyes Publications, Park Ridge, NJ (1993).

24. AC Bailey and B Yates, *J. Appl. Phys.* **41**, 5088 (1970).

25. A Thess et al., *Science* **273**, 483 (1996).

26. SM Sharma et al., *Phys. Rev. B* **63**, 205417 (2001).

27. SM Clark, K-J Jeon, J-Y Chen, C-S Yoo, *Solid State Commun.* **154**, 15 (2013).

28. VD Blank, SG Buga, GA Dubitsky, NR Serebryanaya, MYu Popov and B Sundqvist, *Carbon* **36**, 319 (1998).

29. C Journet et al., *Nature* **388**, 756 (1997).

30. J-F Colomer, L Henrard, G Van Tendeloo, A Lucas and P Lambin, *J. Mater. Chem.* **14**, 603 (2004).

31. O Lourie, DM Cox and HD Wagner, *Phys. Rev. Lett.* **81**, 1638 (1998).

Preparation and Processing of Graphene and SWCNTs

9.1 Motivation

A wide variety of methods have been utilized to prepare samples of graphene and carbon nanotubes (both single-walled and multi-walled), leading to the existence of a very wide range of materials described as "graphene" and "carbon nanotubes". A knowledge of the different synthesis methods is necessary to understand why different graphene and SWCNT samples have the exhibited properties, and to understand which synthesis methods can produce suitable samples for a given experiment or application. The scalability and cost of the different production methods also varies widely.

Post-growth processing of graphene and SWCNTs is also described in this chapter. In particular, we describe how it is possible to transfer the graphene to different substrates after growth and to separate individual SWCNTs according to diameter and (n,m) chiral indices.

9.2 Graphitization of carbon materials via heat treatment

Before discussing the specific methods to produce graphene and SWCNTs, it is appropriate to understand what happens when we heat up an amorphous or disordered carbon material in vacuum or an inert atmosphere. At ambient conditions, Bernal (ABAB) stacked graphite is the thermodynamically stable phase of carbon, and this remains the case as temperature is increased at ambient pressure or vacuum. As we increase temperature, we give the atoms in the disordered carbon sample more and more thermal energy to arrange themselves into the thermodynamically stable phase, and we can continue doing this until we reach the sublimation point at 3915 K. As we heat the carbon, we expect the degree of disorder to decrease, but it turns out that the degree of disorder as a function of temperature is not a linear function. Instead, there are various temperature ranges associated with different degrees of structural order ([12] of Chapter 1 and [1,2]) (Figure 9.1).

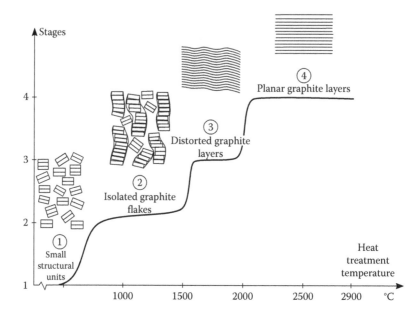

FIGURE 9.1

Schematic diagram of the four stages in the heat treatment of disordered carbon. Not until temperatures of over 2000°C is a defect-free crystal of graphite produced by the heat treatment process. (Reprinted from *Carbon*, 22, Oberlin, A., 521, Copyright 1984, with permission from Elsevier.)

When we synthesize graphene and SWCNT in a bottom-up process such as chemical vapour deposition, we need to look at where the process temperature lies on this graph. When we synthesize graphene by a top-down process such as mechanical exfoliation, we need to look at the production temperature of the graphite sample we begin with. In both cases, by doing so we can gain some idea of the degree of crystalline quality to expect in the products. The mechanical exfoliation (next section) and liquid-phase exfoliation (Section 9.3.2) methods of graphene production are normally performed using highly oriented pyrolytic graphite or natural vein graphite. Highly oriented pyrolytic graphite is produced by heat treatment at 2700°C [3] and natural graphite has been subjected to a similar temperature at some point in time in the upper layers of the earth's mantle, so in both cases it is possible to obtain graphite samples with very few defects.

9.3   Preparation of graphene

9.3.1   Mechanical exfoliation

The mechanical exfoliation, or "scotch tape" method, was the method first used to isolate mono-layer, bi-layer and few-layer graphene in 2004 ([24] of Chapter 1). The method

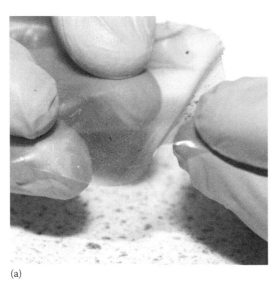

(a)

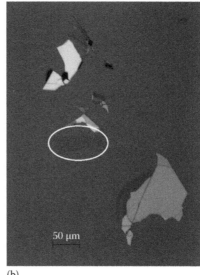

50 μm

(b)

FIGURE 9.2
(a) Graphene and thin graphite samples adhered to adhesive tape during the mechanical exfoliation process for graphene production. (b) A mono-layer graphene flake (circled) visible (just) in an optical microscope on a silicon wafer covered in 300 nm thick SiO_2 layer.

exploits the weakness of the van der Waals bonding between the atomic layers of graphite flakes. A crystal of graphite is peeled apart many times with adhesive tape (Figure 9.2) until, in some locations, there are flakes of mono-layer, bi-layer and few-layer graphene left adhered to the tape. The adhesive tape, with flakes attached, is then pressed against a substrate and peeled away. In some cases the graphene crystals remain attached to the substrate, along with much larger quantities of few-layer graphene and thin graphite crystals.

To exfoliate graphene crystals with the extremely low defect density required to observe the effects discussed in previous chapters (for instance, the observation of the quantum Hall effect at ambient temperature), it is necessary to begin the process with extremely high-quality natural vein graphite or highly oriented pyrolytic graphite. The adhesive tape used must be of a suitably clean scientific grade in order to avoid contaminants on the graphene surface.

The mono-layer graphene flakes produced by this method occupy only a tiny part of the area of the substrate, and are mixed in with a much larger number of bi-layer and few-layer graphene flakes, as well as a lot of thin graphite crystals. Therefore, it is crucial to have a method to find the mono-layer graphene flakes. The most common method is also the simplest. It is to use as the substrate a silicon wafer with a 300 or 90 nm thick layer of SiO_2 thermally deposited on the surface. This allows even a mono-layer graphene flake to be visible with the naked eye in an optical microscope due to the tiny change in the refractive index of the path taken by the light [4]. Even a tiny change in SiO$_2$ layer thickness from 300 to 315 nm can make graphene invisible. The quickest and most reliable method to check the suitability of the oxide layer thickness is through its colour as this varies rapidly as a function of thickness. A 300 nm thick layer is pinkish purple in colour. Figure 9.2 shows a mono-layer graphene sample made visible by this method, and the reader is referred to colour images in the scientific literature ([24] of Chapter 1 and [4,5]).

In addition, the effective observation of graphene requires a good-quality optical microscope with illumination by white light arriving through the objective lens. Whilst not quite as simple and low-tech as articles in the popular media would have us believe, there can be no doubt that the mechanical exfoliation technique has provided a relatively low-cost method to produce exceptionally high-quality graphene samples on a suitable substrate for measurements of electronic properties and other scientific research.

However, the extremely small throughput and time-consuming procedure to identify the graphene samples ensures that this technique is completely unsuitable for commercial production of graphene samples. For this, we must look primarily to the liquid-phase exfoliation and chemical vapour deposition techniques described in the next two sections.

9.3.2 Liquid-phase exfoliation

The mechanical exfoliation method produces extremely small quantities of high-quality mono-layer graphene alongside much larger quantities of few-layer graphene and thin graphite crystals and the extremely small quantity of graphene produced renders it unsuitable for commercial applications. As an alternative, the liquid-phase exfoliation method has been developed. This method involves immersing graphite flakes in a solvent and subjecting the suspension to ultrasound, in order to simply shake the graphite apart into graphene flakes. The technique was first reported simultaneously by two groups in 2008 [6,7] and has been utilized extensively in graphene research since then. It can produce far greater quantities of graphene than the mechanical exfoliation process.

To begin, the liquid-phase exfoliation process consists of subjecting a suspension of graphite flakes in a solvent to ultrasound. The solvent is chosen such that the energetic cost of exfoliating the graphite flakes into graphene flakes surrounded by solvent molecules is as small as possible, a choice that can be understood in terms of the surface energy [8] of the graphite* and the solvent. The energetic cost of exfoliation

---

* Here we define surface energy as the post per unit surface area of separating adjacent graphene sheets, not ripping individual graphene sheets.

is the cost of separating out the solvent molecules, plus the cost of separating out the graphene sheets from the graphite crystal, minus the gain in energy when the solvent molecules are allowed to bond to the surface of the individual graphene layers (via van der Waals bonding only). If the solvent surface energy is too large then the energetic cost of separating the solvent molecules is too high for the process to occur. If, on the other hand, the solvent surface energy is too low, then not enough energy is saved when the solvent bonds to the graphene flakes.

Applying this logic in a more rigorous manner, it can be shown that the energetic cost of exfoliating the graphite flakes to form graphene is at a minimum when the surface energy of the graphite and the solvent is equal [7]. The solvent–graphene surface energies can be modified by the use of surfactants which coat the graphene. The addition of surfactants allows the use of water as a solvent.

After sonication, a small proportion of the original graphite sample is exfoliated into mono-layer, bi-layer and few-layer graphene samples, while most remains in the form of thicker graphite flakes suspended in the solvent. The thicker flakes can then be removed using centrifugation to leave a suspension of mono-layer, bi-layer and few-layer graphene samples in the solvent.

A typical yield of mono-layer graphene as a percentage of the total mass of the original graphite crystal is 1 wt.%. The yield as a percentage of the total mass of the mono-layer, bi-layer and few-layer graphene samples left over following centrifugation is 12 wt.%, or 25% in terms of the proportion of flakes which are mono-layer. Figure 9.3 shows a typical distribution of flake thicknesses in a graphene sample produced by the liquid-phase exfoliation process, following centrifugation. It is worth noting, however, that the graphite sediment removed during the centrifugation process can be subjected to further attempts at exfoliation to increase this yield.

Following the sonication and centrifugation processes the graphene flakes produced can be deposited onto a substrate such as a TEM grid or processed further in solution.

To achieve the yields mentioned earlier, the starting graphite flakes are sonicated for about 30 min. Using longer sonication times (up to 460 h) can significantly increase the yield of mono-layer graphene samples in this process, but at the expense of introducing a much greater number of defects into the graphene sheets [9]. However, some ripping of the individual graphene sheets does occur even for the lowest sonication times used [10] and the method yields mono-layer graphene samples of smaller size and lower crystalline quality than the mechanical exfoliation process [11].

In summary, the liquid-phase exfoliation process described here is important due to its scalability to produce large quantities of graphene for commercial use. However, it suffers

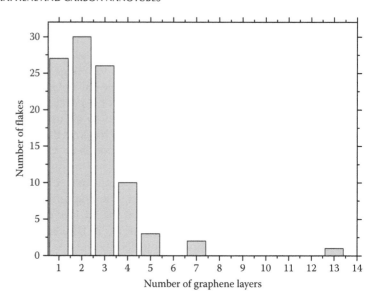

FIGURE 9.3
Histogram showing thickness distribution of graphene flakes produced using the liquid-phase exfoliation process following centrifugation. (Reprinted by permission from Macmillan Publishers Ltd. *Nat. Nanotechnol.*, Hernandez, Y. et al., 3, 563, Copyright 2008.)

from the crucial disadvantage that (just as for the mechanical exfoliation process) the mono-layer graphene samples produced are mixed in with a large quantity of few-layer graphene and thin graphite samples. In addition, the sample quality is not as high as that achieved using the mechanical exfoliation process.

Whilst the mechanical and liquid-phase exfoliation processes are both top-down processes, efforts have been made since the beginning of graphene research to produce graphene in a bottom-up process instead. The aim is to produce large single crystals of graphene on substrates suitable for further study in a process scalable for commercial use. The mechanical and liquid-phase exfoliation processes both fail to achieve this. The first bottom-up process developed for graphene synthesis was epitaxial growth (Section 9.3.3) and the second (more widely used process) is chemical vapour deposition (Section 9.3.4).

9.3.3   Epitaxial growth

The principal substrate used for the epitaxial growth of graphene is silicon carbide, which has been utilized for the epitaxial growth of mono-layer graphene since 2006 [12]. It has been utilized for the epitaxial growth of few-layer graphene since 2004 [13], producing samples in which quantized cyclotron motion of electrons in a magnetic field was observed simultaneously to its observation in mono-layer graphene by Novoselov et al. ([24] of Chapter 1).

Epitaxial growth of graphene is typically performed on the silicon-terminated (0001) face of 6H-SiC by annealing under vacuum or argon at a temperature of 1250°C–2000°C [13–15]. Annealing at ambient pressure under argon was a later (2008) development which was found to significantly increase the grain size of the graphene flakes [15,16]. In either case, the graphene is grown from carbon atoms which come from the top layer of the SiC when it decomposes at high temperature. The silicon atoms are lost to the atmosphere via sublimation and the carbon atoms are then free to form graphene. The growth can be made self-limiting as, once the graphene layer is formed, further sublimation of the silicon cannot occur. Epitaxial graphene growth has also been achieved on boron carbide [14,17]. In this case, due to the similar atomic size of boron and carbon, the graphene is strongly doped with boron atoms.

The crystalline quality achieved by the best epitaxially grown graphene samples is very high indeed; the quantum Hall effect has been observed in such samples [16]. However, there are drawbacks to the epitaxial growth technique. In particular the fact that, as is usually the case in epitaxial growth, the graphene layer interacts strongly with the substrate. The orientation of the graphene layer is determined by the orientation of the substrate lattice [14] and the substrate can even cause the appearance of a superlattice in the graphene [17].

As a consequence of this, it has been shown that graphene grown by epitaxial growth on SiC is under significant compressive strain. The growth takes place at high temperature. However, upon cooling the SiC substrate contracts whilst the graphene (at least as it approaches ambient temperature) is trying to expand due to its negative thermal expansion coefficient in the in-plane direction (as discussed in Chapter 10). Raman spectroscopy indicates that the in-plane compressive strain induced in the graphene by this process is equivalent to several gigapascals of hydrostatic pressure [18–20].

The high cost of SiC wafers is an additional barrier to the commercial use of graphene grown by this method [21,22]. For these reasons, despite the promising early results achieved using the epitaxial growth process, chemical vapour deposition is now the more widely used technique for large-scale graphene growth.

9.3.4   Chemical vapour deposition (CVD)

The method of graphene growth most likely to prove suitable for commercial applications is chemical vapour deposition (CVD) on a transition metal substrate. A variety of precursor

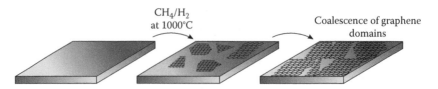

FIGURE 9.4

Schematic diagram illustrating stages in the growth of graphene by CVD on copper. When growth commences, a number of small crystallites form simultaneously at different locations on the substrate. Upon completion of growth (the entire substrate covered in a mono-layer of graphene) defects are present at the boundaries between the crystallites.

gases and experimental conditions have been utilized, but the most common method is vacuum* CVD at approximately 1000°C with methane as the precursor gas and hydrogen also present. The process is catalytic because the transition metal substrate catalyses the decomposition of the precursor gas to produce the carbon which forms the graphene layer.

Copper [23] is the most commonly used substrate, followed by nickel, but cobalt, ruthenium, palladium, rhenium, iridium and platinum have also been used [21]. In some cases the lattice of the CVD-grown graphene is commensurate with that of the substrate, but this is not observed so commonly as in epitaxial growth. In the cases of growth on the Co(001) and Ni(001) surfaces the graphene and substrate lattices are commensurate, whilst in the cases of Pt(111), Pd(111), Ru(111) and Ir(111) the lattices are incommensurate [21].

The crystalline quality of the graphene grown by CVD is lower than that of graphene produced by the mechanical exfoliation method. There are two explanations for this. Firstly, when methane is introduced to the reactor the growth process commences simultaneously in various different locations on the substrate. The graphene sample therefore consists of a large number of small crystals instead of one single crystal, as illustrated in Figure 9.4.

Secondly, the growth temperature is relatively low. Growth typically takes place at a temperature of 1000°C whilst the HOPG and natural graphite used as the starting materials for the mechanical and liquid-phase exfoliation processes have been subjected to temperatures above 2000°C in the factory or in the earth's interior. Referring back to Figure 9.1, we can expect that this growth temperature will produce samples with a significant amount of defects. Unfortunately, in the case of copper, a higher growth temperature would result in the substrate melting. Despite these limitations the quantum Hall effect has been observed in graphene samples grown

---

* Mono-layer graphene has, however, also been grown using ambient pressure CVD [32].

by CVD on both copper [24] and nickel [25], indicating that CVD-grown graphene is certainly of adequate crystalline quality for electronics applications. Graphene grain sizes of the order of 100 µm are routinely achieved [26], and a single graphene crystal can grow across grain boundaries in the substrate.

Copper is the most widely used substrate because of the high proportion of the substrate which can be covered in mono-layer graphene (as opposed to bi-layer graphene or few-layer graphene). This is due to the growth of graphene on copper being a self-limiting process. The growth process commences with the copper substrate being annealed at high temperature in hydrogen. When copper is subjected to ambient conditions an oxide layer (CuO, Cu_2O) rapidly forms on the surface [27]. When annealing takes place the hydrogen reacts with this copper oxide surface layer, leaving pristine copper exposed to methane molecules when they are introduced to the reaction chamber.

Pristine copper does not react with carbon (no copper carbide phases are known [28,29]), and the solubility of carbon in copper is extremely low (0.001–0.008 wt.%) [21]. Hence, virtually no carbon is absorbed into the copper substrate. The sole source of carbon for growth is carbon atoms resting on the surface of the copper after the copper has acted as a catalyst for the decomposition of the methane molecules. Once a single atomic layer of graphene has grown on the copper surface any more methane molecules that arrive do not have access to the copper to catalyse their decomposition [26,30]. Hence, even for long growth times, the thickness of the graphene produced is limited to one layer.

Graphene samples grown by CVD on copper are over 95% mono-layer [26,30]. Generally, growth on nickel leads to multilayer graphene but graphene samples 87%–91% mono-layer have been grown on single crystal nickel by utilizing a very rapid cooling rate [26,30].

The interaction between the copper substrate and the graphene sample after growth is also extremely weak, compared to that between graphene and other transition metals (or silicon carbide for that matter). This is due to the electronic structure of copper ($[Ar]3d^{10}4s^1$). The 3d shell is full, and hence the only bonding possible between graphene and copper is some charge transfer from the graphene π orbitals to the half-full 4s orbital in the copper.

The weakness of this interaction is reflected in the behaviour of the system upon cooling after growth. As discussed earlier, graphene grown epitaxially on SiC is left under considerable compressive strain after cooling due to the different thermal expansion coefficients of graphene and SiC. When graphene is grown by CVD on copper, there is also a substantial mismatch in thermal expansion coefficient between the graphene crystal and the substrate – the substrate compresses significantly upon cooling and the graphene crystal does not. However, due to the interaction between graphene and copper being much weaker, the graphene grown by CVD on copper is not left under strain. The Raman peaks of the graphene are in the same position as those of (unstrained) graphene produced on SiO_2 using the mechanical exfoliation process. Instead, substantial wrinkles form in the graphene crystal.

Finally, it is essential to understand the key role played by the hydrogen gas present during the CVD process. In a typical CVD growth process for graphene the copper substrate in the CVD furnace is first heated up under low hydrogen pressure (~0.1 Torr) to above 1000°C, before methane is introduced to initiate growth of graphene. During the heating stage, the hydrogen is responsible for the removal of the oxide layer from the copper surface as discussed earlier.

However, the hydrogen flow is continued during the growth period. During this period the hydrogen plays a different, and every bit as important, role. On the copper surface, the methane gas molecules decompose to free hydrogen and carbon, but the free hydrogen and carbon can also recombine to form methane. So an equilibrium state exists where both reactions are taking place simultaneously (Equation 9.1) due to le Chatelier's principle:

$$CH_4 \leftrightarrow C + 2H_2 \tag{9.1}$$

Adding hydrogen separately causes the equilibrium state in this reaction to shift towards a higher concentration of reactants (methane) and a lower concentration of products (hydrogen and free carbon). Without this addition of hydrogen, too much free carbon is deposited on the substrate too quickly and the growth of graphene is no longer self-limiting to one layer [31]. On the other hand, if too much hydrogen is introduced to the reaction chamber the hydrogen bonds to the carbon atoms in the graphene crystal [32], most likely creating a form of highly disordered hydrogenated graphene. This is evident from the appearance of the first-order Raman D peak in such samples. The reader is referred to Reference 33 for a more detailed discussion of the role of hydrogen in the CVD growth of graphene.

9.3.5   Transfer of graphene to different substrates

It is often necessary to obtain a graphene sample on a substrate tailored to a specific experiment or application. For instance, to perform an electron diffraction experiment on graphene, it is necessary to utilize a sample mounted on a substrate with an aperture, and to study the electric field effect in graphene it is necessary to utilize a sample mounted on an insulating substrate on which contacts can easily be deposited. Fortunately, it is relatively straightforward to transfer graphene to different substrates following synthesis/exfoliation.

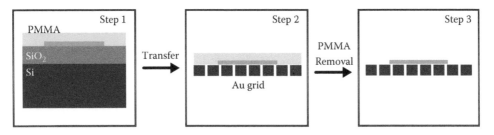

FIGURE 9.5

Schematic diagram illustrating procedure for transfer of graphene between different substrates utilizing poly(methyl methacrylate) (PMMA), in this case from SiO_2 to an Au grid. (Adapted from Nair, R.R. et al.: *Small*. 2010. 6. 2877. Copyright Wiley-VCH Verlag GmbH & Co. KGaA. Reproduced with permission.)

The most common procedure to transfer graphene is to cover the graphene sample in a thin layer of the glass poly(methyl methacrylate) (PMMA). The original substrate is then etched away. There are a number of reagents which can etch away the substrates on which graphene is typically synthesized/exfoliated whilst leaving the graphene layer and PMMA intact, for instance, potassium hydroxide solution can etch away an SiO_2 substrate in this procedure. The PMMA layer (with graphene attached) is then placed on the new substrate, after which the PMMA is removed by dissolving it in acetone. The graphene is then left on the new substrate in the desired location. This procedure (described schematically in Figure 9.5) can be performed routinely on graphene samples produced by the mechanical exfoliation method on SiO_2 (for instance, Reference 34) and by CVD on copper (for instance, Reference 2 of Chapter 8).

9.4   Growth and processing of SWCNTs

A variety of methods have been developed to grow SWCNTs since their discovery in 1993. In the decade following their discovery, the principal methods utilized for the growth of SWCNTs were the carbon arc method (Section 9.4.1), the laser ablation method (Section 9.4.2), the catalytic (Section 9.4.3) and plasma-enhanced (Section 9.4.4) CVD methods. Whilst these methods offer some control over the diameter of the SWCNTs grown, they do not offer control over the chirality. In most cases, these methods result in the growth of SWCNTs in bundles. However, it is now possible to separate out the individual SWCNTs from the bundles using relatively simple and inexpensive methods (Section 9.4.6).

The synthesis methods used most widely nowadays, and displaying the best potential for more widespread commercial use in the future, are the chemical vapour deposition methods (catalytic CVD and plasma-enhanced CVD). The arc discharge method suffers from the disadvantage that the SWCNT yield is very low compared to the large quantity of impurities formed in the process, and both the arc discharge and laser ablation methods suffer from the disadvantage that they require solid graphite as a starting material. This prevents the production of very large quantities of SWCNTs since the cost of obtaining the required carbon in the form of graphite rather than hydrocarbon gas is prohibitive.

For some applications (for instance, the use of SWCNTs in composite materials), chirality-specific synthesis is not necessary and the growth methods developed in the decade following the discovery of SWCNTs in 1993 are adequate. However, as discussed in Chapter 5, the electronic dispersion relation of SWCNTs depends strongly on chirality. Tubes of very similar diameter can be metallic, or exhibit relatively large electronic bandgaps, depending on the exact value of the (n,m) assignment. For the realization of the potential applications of SWCNTs in electronics and optoelectronics, it is therefore essential to be able to control the chirality of the tubes in a sample, either by controlling the chirality during synthesis or by sorting the tubes by chirality following synthesis. Several breakthroughs in both methodologies have been made over the past few years ([27] of Chapter 7 and [35,36]) and SWCNT samples in which over 50% of the tubes have the same (n,m) assignment are available commercially [37]. These recent developments will be reviewed in Sections 9.4.7 and 9.4.8.

9.4.1   Carbon arc

Our discussion of the carbon arc method is included here mainly for historical interest. The carbon arc method was the method utilized in the first synthesis of SWCNTs ([14,15] of Chapter 1) and was widely used prior to that in the synthesis of fullerenes [38] and multi-walled carbon nanotubes ([13] of Chapter 1).

The carbon arc method involves the use of a spark between two graphite electrodes to generate conditions suitable for nanotube growth. Typical conditions are 5–20 nm diameter electrodes with a separation of 1 mm. A DC current of ~100 A flows between the electrodes. The temperatures achieved during the growth process can be as high as 3000°C. The system is usually operated in an inert atmosphere of helium at ~500 Torr. For the growth of SWCNTs (as opposed to multi-walled carbon nanotubes) the use of a catalyst is necessary [39]. Rare earth catalysts such as yttrium and gadolinium and transition metals such as cobalt, nickel and iron are commonly used ([12] of Chapter 1). In all cases, large quantities of impurities are generated by this process, including multi-walled carbon nanotubes, fullerenes, nanometre size carbide particles and metal clusters encapsulated in graphene layers ([14,15] of Chapter 1). The SWCNTs produced are either formed in ropes, or surrounded by impurities.

For this reason the carbon arc method has been largely superseded by other, cleaner, methods of SWCNT synthesis.

9.4.2   Laser ablation

This method for growth of SWCNTs uses a high-power laser beam (instead of an electric arc) to evaporate a graphite target and create the high temperatures necessary for nanotube formation. As with the carbon arc method, the graphite target contains ~1% mixed transition metal catalyst, but in all cases the SWCNTs are produced in ropes ([12] of Chapter 1). High yields of 70%–90% conversion of graphite to SWCNTs have been reported ([25] of Chapter 8), in contrast with only ~20% yield with the carbon arc method ([29] of Chapter 8). Figure 9.6 shows a schematic of a typical apparatus for growth of SWCNTs by laser ablation. The temperature generated by the laser beam is extremely high – to evaporate the graphite target the temperature must reach at least the sublimation point of carbon, 3915 K.

9.4.3   Catalytic CVD

Catalytic chemical vapour deposition (CCVD), in common with plasma-enhanced chemical vapour deposition (PECVD – see Section 9.4.4), differs from the arc discharge and laser ablation methods in that hydrocarbon gas is used as a source of carbon.

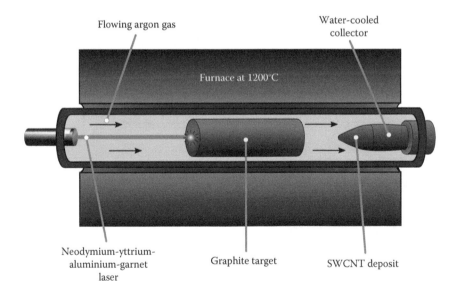

FIGURE 9.6
Typical apparatus for growth of SWCNTs by laser ablation. A graphite target is in inert argon atmosphere at 1200°C. A pulsed Nd-YAG laser evaporates the target and SWCNTs are swept onto the water-cooled collector outside the furnace by flowing argon gas ([12] of Chapter 1 and [40]).

Methane, ethane, ethylene and acetylene are commonly used. The gas thermally decomposes into dimers (C_2) and trimers (C_3). Under the experimental conditions utilized in the CCVD process, the gas does not decompose unless the catalyst particles are present to facilitate the process [41,42]. Similarly to the synthesis of graphene by CVD (Section 9.3.4) hydrogen is added so that the equilibrium in the decomposition reaction will be shifted towards the reactant according to Le Chatelier's principle (Equation 9.1), and the deposition of amorphous carbon will be prevented.

Due to having a gaseous source of carbon, it is possible to grow SWCNTs on a substrate using CVD. In this case the transition metal catalyst is simply deposited on the substrate (often silicon) as nanoparticles before growth of the SWCNTs commences. This is a convenient method of producing SWCNTs to study using techniques such as Raman spectroscopy or scanning electron microscopy. A number of techniques are used to deposit the catalyst on the substrate, and growth temperatures range from 600°C to 900°C in an inert atmosphere. Reference 16 of Chapter 8 describes a typical CCVD apparatus. A very high yield of SWCNTs (in the sense that only ropes of SWCNTs are grown on the silicon substrate) is possible although it is inevitable that the yield in terms of the proportion of carbon fed into the furnace as gas that ends up as nanotubes is very low. CCVD does however generally result in poorer quality SWCNTs than laser ablation and PECVD (i.e. more defects) due to the lower growth temperature (650°C–1000°C). Figure 9.7 is a schematic diagram of typical apparatus for the production of SWCNTs by CCVD.

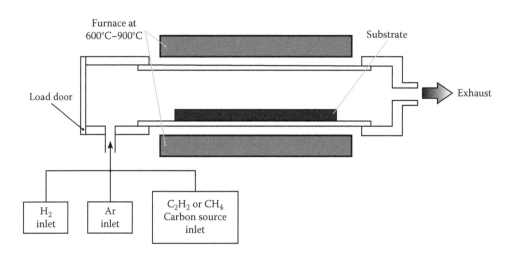

FIGURE 9.7

Schematic of a typical apparatus for growth of SWCNTs by CCVD. SWCNTs are grown following thermal decomposition of hydrocarbon gas on the surfaces of transition metal catalyst nanoparticles resting on a substrate.

The HiPCO® [43] and CoMoCAT® [44,45] processes used to produce commercially available SWCNTs with relatively narrow diameter distribution are both variants on the CCVD process. To produce HiPCO SWCNTs the catalyst is injected in gaseous form. At about 1000°C and 30 atmospheres pressure CO gas is mixed with gaseous $Fe(CO)_5$. The decomposition of this gas results in the formation of metallic nanoparticles that act as catalyst for growth of SWCNTs in the gas phase. In the case of the CoMoCAT SWCNTs, the catalyst nanoparticles are injected into a vertical reactor in freestanding form and held in the reactor by the upward flow of CO gas.

9.4.4   Plasma-enhanced CVD (PECVD)

In plasma-enhanced chemical vapour deposition (PECVD) the kinetic energy to enable the production of free carbon and the growth of SWCNTs is provided partially by heating the entire furnace (as in CCVD) and partially by a plasma generated over the substrate on which the SWCNTs are grown. Most commonly, the plasma is produced by a direct current electric discharge, or radio frequency induction [46]. The PECVD process offers two advantages over CCVD. Firstly, the higher temperature in the immediate vicinity of the substrate allows the growth of SWCNTs with fewer defects. Secondly, the use of a plasma enables the remainder of the furnace to be kept at a lower temperature than in CCVD. This has the potential to reduce the costs of commercial production.

The large variety of specific recipes used for PECVD have been reviewed extensively in the literature [39,46]. Generally, hydrocarbon gas is used as the carbon source and transition metal catalyst particles are used to initiate SWCNT growth, just as for CCVD. However, due to the higher growth temperature, it is possible for the hydrocarbon gas to decompose without the aid of the catalyst [46]. PECVD has even been used to grow aligned ropes of nanotubes protruding from the substrate [47].

9.4.5   Diameter and chirality distribution of SWCNT grown using arc discharge, laser ablation, CCVD and PECVD methods

All methods for SWCNT synthesis discussed so far suffer from a fundamental limitation: They provide no control over the chirality of the SWCNTs synthesized, and only limited control over the diameter. We will recall our discussion in Chapter 5 of the electronic properties of SWCNTs: As a result of the quantum confinement of the electrons in the direction around the circumference of the SWCNT the electronic dispersion relation depends strongly on the specific (n,m) assignment: The (6,0) tube is metallic, whilst the (6,1) and (5,1) nanotubes are semiconducting.

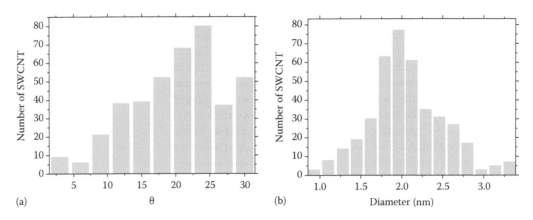

(a)   (b)

FIGURE 9.8

Statistical analysis of the chiral angle (a) and diameter (b) distributions of a sample of SWCNTs grown in a CCVD process. The sample size was 402 SWCNTs, of which 240 were semiconducting. (Reprinted by permission from Macmillan Publishers Ltd. *Nat. Nanotechnol.*, Liu, K. et al., 8, 917, Copyright 2013 [Chapter 5].)

Figure 9.8 shows the results of a statistical analysis of the chiral angle, and metallic or semiconducting nature, of SWCNTs grown in a single CCVD growth run ([18] of Chapter 5). The SWCNTs studied have a wide diameter distribution and the proportion of SWCNT which are metallic and semiconducting is roughly what we expect if we have no control over this parameter: About two-thirds are semiconducting whilst about one-third are metallic. The large proportion of SWCNTs having a large chiral angle is not an accident. It is what is typically observed in SWCNT growth. It is believed to result from a play-off between speed of growth kinetics and energetic favourability of achiral growth at the SWCNT–catalyst interface [48].

Whilst direct control over the chiral angle is not possible using the methods described thus far, some control over the diameter of the synthesized SWCNTs is possible. This is achieved through careful selection of growth parameters, in particular the choice of catalyst particles. The HiPCO process discussed earlier produces SWCNTs with diameters from 0.8 to 1.4 nm and a mean diameter of 1.1 nm [49] whilst the more recently developed CoMoCAT process can produce SWCNTs with a variety of different, and relatively narrow, diameter distributions: 1.7–1.3, 0.7–0.9 and 0.7–1.1 nm [37]. In this case, the diameter distribution is sufficiently narrow that a relatively small range of (n,m) assignments lie within it. The proportion of tubes with semiconducting or metallic character can, in this case, vary significantly from what would be expected statistically from all possible (n,m) assignments. However, a mixture of SWCNTs with some different (n,m) assignments is always present.

9.4.6 Purification of SWCNT samples and debundling of SWCNT

When SWCNT growth takes place, the tubes are produced in bundles (unless there is a particularly low yield of SWCNTs or large amounts of impurities present to prevent bundle formation). These bundles are held together by the van der Waals interaction, the same interaction that bonds layers of graphene to each other to form graphite. In addition, all of the commonly utilized methods to produce SWCNTs involve growth of SWCNTs catalysed by (typically) transition metal nanoparticles. These particles are thus always present as impurities in synthesized samples of SWCNTs. They can be removed by dissolving them using acid (e.g. HCl/HF) [45,49,50].

Following this process, it is possible to perform further processing on the sample to break up the bundles and obtain individual SWCNTs. Since SWCNTs are just one atom thick, the environment of an individual SWCNT (the other SWCNTs in the bundle) has a significant effect on the observed properties of the individual SWCNT, and the technology to remove the SWCNTs from the bundle is of great importance. For instance, the observation of luminescence from the semiconducting SWCNTs in a bundle is prevented by non-radiative recombination between the electrons and holes facilitated by the metallic SWCNTs ([9,10] of Chapter 5).

The development of non-destructive methods to reliably separate SWCNTs from the bundle has therefore been a priority since their discovery in the early 1990s. The method most widely used is to immerse the SWCNTs in a solvent and subject the sample to vigorous ultrasound ([9] of Chapter 5). Unsurprisingly, the methodology utilized is the same as that to perform liquid-phase exfoliation of graphene as discussed in Section 9.3.2. Analogously to the exfoliation of graphene, the procedure works when the surface energy of the solvent is similar to that with which the SWCNTs are bound together in the bundle [51].

The procedure produces individual SWCNTs and very small bundles of SWCNTs in which the SWCNTs are surrounded by molecules of the solvent utilized in the exfoliation, bonded (in most cases) non-covalently to the SWCNTs. This allows (for instance) the observation of strong photoluminescence from individual SWCNTs (see Chapter 5). In addition, individual SWCNTs can be deposited onto a substrate from these solutions [51].

Generally, the debundling procedure does not introduce a large quantity of defects in otherwise pristine SWCNTs. However, it is important to note that (just as the liquid-phase exfoliation process for graphene described earlier results in occasional ripping of the graphene sheets) the sonication part of the debundling procedure does result in occasional cutting of the SWCNTs and hence a reduction in their average length ([9] of Chapter 5).

9.4.7   Selection of SWCNTs according to diameter and chirality

A substantial research effort in recent years has focussed on separating out SWCNTs according to diameter and chirality following synthesis. This is achieved through the separation of the SWCNT bundle into individual SWCNTs in solution as described in the previous section, followed by centrifugation. The centrifugation is intended to separate out the sample of suspended SWCNTs according to their mass, and hence diameter. As far as separation according to chirality ((n,m) indices) is concerned, the discovery that some materials will bond preferably to metallic or semiconducting SWCNTs, or even to SWCNTs with a certain (n,m) assignment [52,53], has proved immensely helpful. The preferential bonding to certain SWCNTs alters the mass of these SWCNTs and hence allows separation using centrifugation or other methods.

Advances in this field over the past decade [52,54,55] have led to the ability to isolate SWCNTs with specific (n,m) assignments ranging from (7,3) to (10,5) with purities of up to 90% utilizing the ability of DNA molecules [56] and allyl dextran based gel [55] to bond preferentially to SWCNTs with specific (n,m) assignments.

9.4.8   Chirality-specific synthesis of SWCNTs

Separate from the research efforts over the course of the last few years to select SWCNTs according to diameter and chirality after growth, attention has turned to the specific possibility of choosing catalyst particles for SWCNT growth which can only catalyse the growth of an SWCNT with a single (n,m) assignment. Three main methods have been attempted to achieve this, as reviewed in detail in Reference 35.

The first method involves using an existing short section of an SWCNT as the catalyst for the CVD growth process. Such short SWCNT samples with (predominantly) a single (n,m) assignment can be produced using the separation techniques described in the previous section then deposited on a substrate. The assembly is then inserted into a CVD reactor and the SWCNTs grow longer from the ends while maintaining their existing chirality [57]. The carbon atoms come from thermal decomposition of a hydrocarbon gas, just as in conventional CVD. The process is analogous to vapour phase epitaxy of three-dimensional semiconductors.

The second method is to use a hydrocarbon molecule as the catalyst particle selected such that under CVD growth conditions, it can fold up on itself (upon dehydrogenation) to form the end cap of a specific SWCNT (Figure 9.9). In particular,

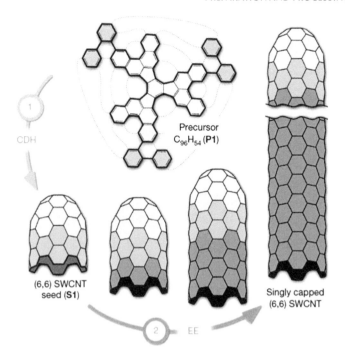

FIGURE 9.9
Cyclo-dehydrogenation (CDH) of the $C_{96}H_{54}$ molecule to form the end cap of a (6,6) SWCNT, followed by growth via epitaxial elongation (EE). (Reprinted by permission from Macmillan Publishers Ltd. *Nature*, Sanchez-Valencia, J.R. et al., 512, 61, Copyright 2014 [Chapter 7].)

the process has been performed utilizing $C_{96}H_{54}$ (to produce (6,6) SWCNTs ([27] of Chapter 7) and $C_{50}H_{10}$ (to produce (5,5) SWCNTs [58]).

The third method involves the use of (inorganic) catalyst nanoparticles prepared to have a specific plane exposed during SWCNT synthesis. This plane is chosen such that the adsorption of carbon atoms on this plane leaves the atoms in the correct positions to begin growth of an SWCNT with a specific (n,m) assignment [36,59]. This method suffers from the disadvantage that it does not produce a success rate close to 100% as the first and second methods do. However, it does have the key advantage that it has been utilized to preferentially grow zigzag SWCNTs with a success rate of 80% [36]. This is in contrast to other SWCNT growth methods (both generic and chirality-specific) which preferentially produce armchair and chiral SWCNTs [48], as discussed earlier.

It is important to note that all of these methods produce a very low yield of SWCNTs compared to the conventional (non-chirality-specific) synthesis methods described in Sections 9.4.1 through 9.4.4.

References

1. A Oberlin, *Carbon* **22**, 521 (1984).

2. J Goma and M Oberlin, *Thin Solid Films* **65**, 221 (1980).

3. LCF Blackman and AR Ubbelohde, *Proc. R. Soc. A* **266**, 20 (1962).

4. P Blake et al., *Appl. Phys. Lett.* **91**, 063124 (2007).

5. KS Novoselov et al., *Proc. Natl. Acad. Sci. USA* **102**, 10451 (2005).

6. P Blake et al., *Nano Lett.* **8**, 1704 (2008).

7. Y Hernandez et al., *Nat. Nanotechnol.* **3**, 563 (2008).

8. R Defay et al., *Surface Tension and Adsorption*, Longmans, London (1966).

9. U Khan, A O'Neill, M Lotya, S De and JN Coleman, *Small* **6**, 864 (2010).

10. M Lotya, PJ King, U Khan, S De and JN Coleman, *ACS Nano* **4**, 3155 (2010).

11. V Nicolosi, M Chhowalla, MG Kanatzidis, MS Strano and JN Coleman, *Science* **340**, 1226419 (2013).

12. C Berger et al., *Science* **312**, 1191 (2006).

13. C Berger et al., *J. Phys. Chem. B* **108**, 19912 (2004).

14. W Norimatsu and M Kusonoki, *Phys. Chem. Chem. Phys.* **16**, 3501 (2014).

15. C Virojanadara et al., *Phys. Rev. B* **78**, 245403 (2008).

16. KV Emtsev et al., *Nat. Mater.* **8**, 203 (2009).

17. W Norimatsu, K Hirata, Y Yamamoto, S Arai and M Kusunoki, *J. Phys. Condens. Mat.* **24**, 314207 (2012).

18. ZH Ni et al., *Phys. Rev. B* **77**, 115416 (2008).

19. J Röhrl, M Hundhausen, KV Emtsev, T Seyller, R Graupner and L Ley, *Appl. Phys. Lett.* **92**, 201918 (2008).

20. JE Proctor, E Gregoryanz, KS Novoselov, M Lotya, JN Coleman and MP Halsall, *Phys. Rev. B* **80**, 073408 (2009).

21. C Mattevi, H Kim and M Chhowalla, *J. Mater. Chem.* **21**, 3324 (2011).

22. KS Novoselov, VI Fal'ko, L Colombo, PR Gellert, MG Schwab and K Kim, *Nature* **490**, 192 (2012).

23. X Li et al., *Science* **324**, 1312 (2009).

24. H Cao et al., *Appl. Phys. Lett.* **96**, 122106 (2010).

25. KS Kim et al., *Nature* **457**, 706 (2009).

26. YI Zhang, L Zhang and C Zhou, *Acc. Chem. Res.* **46**, 2329 (2013).

27. MZ Butt, *J. Mater. Sci. Lett.* **2**, 1 (1983).

28. R McLellan, *Scr. Metall.* **3**, 389 (1969).

29. GA Lopez and EJ Mittemeijer, *Scr. Mater.* **51**, 1 (2004).

30. DAC Brownson and CE Banks, *Phys. Chem. Chem. Phys.* **14**, 8264 (2012).

31. S Bhaviripudi, X Jia, MS Dresselhaus and J Kong, *Nano Lett.* **10**, 4128 (2010).

32. L Gao, W Ren, J Zhao, L-P Ma, Z Chen and H-M Cheng, *Appl. Phys. Lett.* **97**, 183109 (2010).

33. M Losurdo, MM Giangregorio, P Capezzuto and G Bruno, *Phys. Chem. Chem. Phys.* **13**, 20836 (2011).

34. RR Nair et al., *Small* **6**, 2877 (2010).

35. F Zhang, P-X Hou, C Liu, H-M Cheng, *Carbon* **102**, 181 (2016).

36. F Yang et al., *JACS* **137**, 8688 (2015).

37. http://www.sigmaaldrich.com/technical-documents/protocols/materials-science/characterization-and.html.

38. W Krätschmer, LD Lamb, K Fostiropoulos and DR Huffman, *Nature* **347**, 354 (1990).

39. A Govindaraj and CNR Rao, Synthesis, growth mechanism and processing of carbon nanotubes, in *Carbon Nanotechnology*, L Dai (ed.), Elsevier, Amsterdam (2006).

40. BI Yakobson and RE Smalley, *Am. Sci.* **85**(4), 324 (1997).

41. DB Hash and M Meyyappan, *J. Appl. Phys.* **93**, 750 (2003).

42. NR Franklin and H Dai, *Adv. Mater.* **12**, 890 (2000).

43. RE Smalley et al., Gas-phase nucleation and growth of single-wall carbon nanotubes from high pressure CO, United States Patent No. US 6,761,870 (2004).

44. B Kitiyanan, WE Alvarez, JH Harwell and DE Resasco, *Chem. Phys. Lett.* **317**, 497 (2000).

45. G Lolli, L Zhang, L Balzano, N Sakulchaicharoen, Y Tan and DE Resasco, *J. Phys. Chem. B* **110**, 2108 (2006).

46. M Meyyappan, *J. Phys. D: Appl. Phys.* **42**, 213001 (2009).

47. ZF Ren et al., *Science* **282**, 1105 (1998).

48. VI Artyukhov, ES Penev and BI Yakobson, *Nat. Commun.* **5**, 4892 (2014).

49. W Zhou et al., *Chem. Phys. Lett.* **350**, 6 (2001).

50. T-J Park, S Banerjee, T Hemraj-Benny and SS Wong, *J. Mater. Chem.* **16**, 141 (2005).

51. SD Bergin et al., *Adv. Mater.* **20**, 1876 (2008).

52. SA Hodge, MK Bayazit, KS Coleman and MSP Shaffer, *Chem. Soc. Rev.* **41**, 4409 (2012).

53. B Liu, *Small* **9**, 1379 (2013).

54. MS Arnold, AA Green, JF Hulvat, SI Stupp and MC Hersam, *Nat. Nanotechnol.* **1**, 60 (2006).

55. H Liu, D Nishide, T Tanaka and K Kataura, *Nat. Commun.* **2**, 309 (2011).

56. X Tu, S Manohar, A Jagota and M Zheng, *Nature* **460**, 250 (2009).

57. J Liu et al., *Nature Comm.* **3**, 1199 (2012).

58. B Liu et al., *Nano Lett.* **15**, 586 (2014).

59. F Yang et al., *Nature* **510**, 522 (2014).

10 Thermal and Mechanical Properties of Graphene and SWCNTs

The thermal and mechanical properties of graphene and carbon nanotubes are a reflection of their 2D and 1D nature, respectively, and are responsible for much of the excitement surrounding these new materials. In this chapter, we review key mechanical and thermal properties of graphene and SWCNTs, focussing also on the experimental obstacles to making precise measurements of these properties.

10.1 Thermal expansion coefficient of graphene

10.1.1 Theory

Readers will no doubt be familiar with thermal expansion in 3D materials. The vast majority of 3D materials expand when heated, and we describe this expansion mathematically using a parameter called the thermal expansion coefficient β:

$$\beta = \frac{1}{V}\left(\frac{\partial V}{\partial T}\right)_P \tag{10.1}$$

We can rewrite Equation 10.1 as follows*:

$$\beta = \frac{1}{B}\left(\frac{\partial P}{\partial T}\right)_V \tag{10.2}$$

where B is the bulk modulus of the material, given by

$$B = -V\left(\frac{\partial P}{\partial V}\right)_T \tag{10.3}$$

It is important to note that β is not a constant; it varies as a function of temperature as the relationship between volume (V) and temperature (T) is generally not linear. Physically, we can understand that thermal expansion is due to the presence of phonons in the material and this is why β is not a constant. The value of β at any given temperature depends

---

* Utilizing $\left(\frac{\partial V}{\partial T}\right)_P = -\left(\frac{\partial V}{\partial P}\right)_T\left(\frac{\partial P}{\partial T}\right)_V$, obtained from a standard identity in partial derivatives [55].

on the characteristics of the phonons which there is adequate thermal energy $(3/2k_BT)$ available to excite at the sample temperature.

Whilst β is not a constant, we can state with confidence that for a 3D material it is usually positive. $(\partial P/\partial V)_T$ is always negative (A material cannot increase its volume when we put it under more pressure!); hence, B is always positive. $(\partial P/\partial T)_V$ is also usually positive; when we heat a 3D material at constant volume, the atoms or molecules will impact the walls of the container more often. This is true whatever the material is; it can be a fluid, liquid, crystalline solid, amorphous solid or an ideal gas, for instance. Hence, according to Equation 10.2 the thermal expansion coefficient β of a 3D material is usually positive. There are a small number of exceptions [1], the most everyday one being water – which exhibits a negative thermal expansion coefficient from 0°C to 4°C in addition to expanding when it freezes into ice.

But do the earlier arguments have to apply to a 2D or 1D material? It turns out not. Scientists have understood for decades the possibility that 2D and 1D materials could have a negative thermal expansion coefficient – "Thermal properties of chain and layered structures at low temperatures" was published by Lifshitz in the *Soviet Journal of Experimental and Theoretical Physics* in 1952 [2], elucidating this point.

To consider a 2D material, we could substitute A (area) for V in the expressions mentioned earlier. We would start with $\beta = (1/A)(\partial A/\partial T)_P$ and eventually arrive at

$$\beta = \frac{1}{B}\left(\frac{\partial P}{\partial T}\right)_A \tag{10.4}$$

where B is the bulk modulus as defined in two dimensions. Just like its 3D counterpart, B is always positive:

$$B = -A\left(\frac{\partial P}{\partial A}\right)_T \tag{10.5}$$

But does $(\partial P/\partial T)_A$ need to be positive? We can understand in simple conceptual terms why $(\partial P/\partial T)_A$ for a 2D material or $(\partial P/\partial T)_L$ for a 1D material (where L is the length) can easily be negative. Consider (Figure 10.1) a free-standing chain of atoms at temperature sufficiently low as to cause negligible thermal motion. Along the axis of the chain of atoms, we exert a pressure P to constrain the chain to a given length. Now, suppose that we increase the temperature and phonons are excited. If most of the phonons excited are out-of-plane phonons then the motion of the atoms in the out-of-plane axes will tend to pull the ends of the chain inwards, reducing the pressure required to contain the chain to a certain length L. Hence, $(\partial P/\partial T)_L$ for a 1D material can be negative, and a similar argument applies to 2D materials.

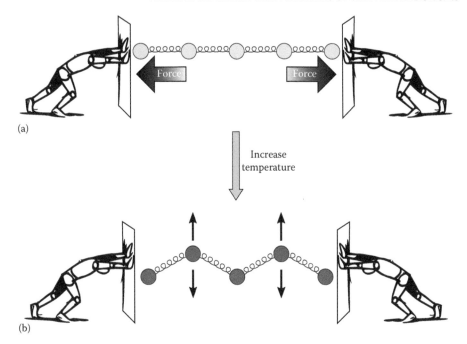

FIGURE 10.1
(a) Chain of atoms at low temperature with negligible thermal motion of the atoms. (b) Chain of atoms following increase in temperature and out-of-plane thermal motion of the atoms. The pressure required to constrain the chain of atoms to a given length is reduced.

To evaluate the thermal expansion coefficient of graphene, we must therefore examine the phonon dispersion relation to discover whether, at a given temperature, we excite mainly in-plane phonons or mainly out-of-plane phonons. The lower-energy section of the phonon dispersion relation of graphene, which we learnt about in Chapter 6, is shown in Figure 10.2. At temperature T, the number of phonons excited from a given dispersion curve in a 2D material is given as follows:

$$N = \int_{k_{min}}^{\infty} \frac{1}{e^{\frac{E(k)}{k_B T}} - 1} \times \frac{k}{2\pi} dk \qquad (10.6)$$

The first term in Equation 10.6 is the Bose–Einstein distribution function accounting for the amount of thermal energy available to excite phonons, and the second term is the 2D density of states as a function of k (Equation 3.19 but ÷2 to account for the fact that phonons are bosons instead of fermions). To account for all possible phonons, we integrate over all k, but due to the finite size of the graphene sheet, there is a maximum wavelength (and hence minimum wavevector k_{min}) of phonon which can be accommodated. In Figure 10.2 the Bose–Einstein distribution is plotted as a function of E for a temperature of 300 K.

By inspection of Figure 10.2, we can see that, certainly at ambient temperature and below, most of the phonons excited

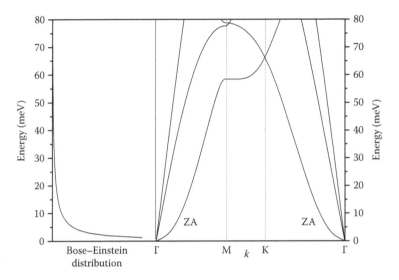

FIGURE 10.2
(Left-hand panel) Bose–Einstein distribution plotted as a function of energy, at 300 K. (Right-hand panel) The low-energy section of the calculated phonon dispersion relation of graphene, also plotted as a function of energy. The out-of-plane acoustic phonon is labelled as ZA.

in a graphene sample are out-of-plane phonons. This is because (Chapter 3) the available quantum states for phonons are equally spaced in k-space so the larger the area of k-space covered by a phonon dispersion curve, the more quantum states will be available for phonons of that variety. As shown by the Bose–Einstein distribution in Figure 10.2 at 300 K, there is only adequate energy to excite phonons with energy up to ca. 30 meV. In this region, the out-of-plane acoustic phonon (labelled ZA) occupies by far the largest region of k-space. Hence most of the phonons excited in a graphene sample at low temperature and at ambient temperature are out-of-plane phonons and free-standing graphene has a negative thermal expansion coefficient under these conditions. This is consistent with the C–C bond length increasing as a function of temperature at all temperatures, as has been suggested [1,3].

10.1.2 Direct measurement of thermal expansion coefficient in graphene

But how can the thermal expansion of a free-standing monolayer graphene sheet be measured? The nearest to a short answer is that it cannot. For a start, a genuinely free-standing graphene sheet is thermodynamically unstable as discussed in previous chapters and proven in Section 10.2. So any real graphene sheet requires support from a substrate, even if part of the sheet is suspended over an aperture in the substrate. Therefore, a real graphene sheet is not free to expand or contract freely.

As if that was not obstacle enough, it has not so far proved feasible to conduct an experiment in which the lattice constant of the mono-layer graphene is directly and accurately measured across a wide temperature range. So how is it possible to measure the thermal expansion coefficient of graphene? The process involves making use of a series of theoretical predictions and experimental assumptions.

An experiment can be performed in which the Raman G peak position* of graphene on a substrate is measured whilst temperature is varied [4–6]. It is assumed that the graphene adheres perfectly to the substrate, so, as temperature is increased the graphene flake is forced to expand with the substrate even though it would rather contract if it was free to do so. We then assume that the experimentally observed shift of the Raman G peak position ($\Delta\omega_E(T)$) can be written as the sum of two shifts: one due directly to the change in temperature ($\Delta\omega_T(T)$) and one due to the change in the in-plane strain on the graphene as temperature is varied ($\Delta\omega_S(T)$):

$$\Delta\omega_E(T) = \Delta\omega_T(T) + \Delta\omega_S(T) \qquad (10.7)$$

$\Delta\omega_T(T)$ is a (relatively) straightforward quantity to predict reliably from first principles [7] and $\Delta\omega_S(T)$ is a function of the parameter we are attempting to determine: The two-dimensional thermal expansion coefficient of graphene† $\alpha_{Gr}(T)$. This is related to $\Delta\omega_S(T)$ by:

$$\Delta\omega_S(T) = \beta_G \int_{T_0}^{T} \left[\alpha_S(T) - \alpha_{Gr}(T)\right] dT \qquad (10.8)$$

where

$\alpha_S(T)$ is the thermal expansion coefficient of the SiO_2 substrate

$\alpha_{Gr}(T)$ is the thermal expansion coefficient of graphene

β_G is the biaxial strain coefficient of the Raman G peak

---

* It is better to use the G peak than the 2D peak because the G peak originates from a non-resonant Raman scattering process. Hence, a theoretical prediction regarding the G peak position dependency on temperature and other parameters can be made without accounting for any changes to the electronic dispersion relation as the graphene contracts/expands.

† It is essential not to get confused between the phonons which cause the thermal expansion and the phonon which we use to measure the thermal expansion. The phonons which cause the thermal expansion are the out-of-plane low-energy acoustic phonons where there is adequate thermal energy available to excite at ambient temperature (Figure 10.2). The phonon which we use to measure the thermal expansion is the in-plane optical phonon at the Γ point (see Figures 6.1, 6.2 and 7.5). There is not enough thermal energy available at room temperature to excite this phonon (so it does not contribute to the thermal expansion until much higher temperatures are reached), but we excite it using laser photons to perform Raman spectroscopy.

β_G can be measured experimentally (Table 10.2) and is described in dimensionless terms using a parameter called the Grüneisen parameter γ_G and the G peak frequency in unstrained graphene at ambient conditions ω_G:

$$\beta_G = -2\omega_G\gamma_G \qquad (10.9)$$

When we substitute Equation 10.8 into Equation 10.7, $\alpha_{Gr}(T)$ is the only unknown variable. We therefore make it the fitting parameter when we compare our theoretically predicted G peak shift as a function of temperature to the experimental data.

Using this methodology, it is possible to fit reasonably well to the experimental data (Figure 10.3).

It is thus possible to calculate the thermal expansion coefficient of graphene as a function of temperature (Figure 10.4). But how reliable is this value? A number of assumptions are made in the process:

1. We accept the results of an *ab initio* prediction [7] of the G band frequency as a function of temperature in free-standing graphene which has not been directly and quantitatively verified.

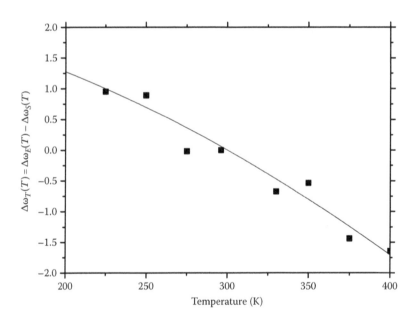

FIGURE 10.3

(Line) Theoretical prediction [7] of the temperature dependence of the Raman G peak in free-standing graphene. (Points) Experimentally observed temperature dependence of the Raman G peak in supported graphene after correction for the strain due to the thermal expansion of the substrate (vs the thermal contraction of graphene). The thermal expansion coefficient of graphene is the fitting parameter used to calculate $\Delta\omega_S(T)$ and hence adjust the experimental data to fit to the theoretical prediction. (Reprinted with permission from Yoon, D., Son, Y.-W. and Cheong, H., *Nano Lett.*, 11, 3227. Copyright 2011 American Chemical Society.)

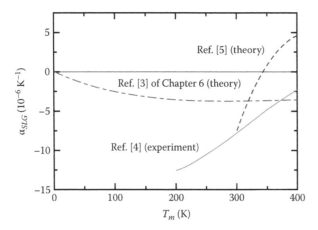

FIGURE 10.4

Comparison of experimental [4] and theoretical ([3] of Chapter 6 and [5]) measurements of the thermal expansion coefficient of graphene. (Reprinted with permission from Yoon, D., Son, Y.-W. and Cheong, H., *Nano Lett.*, 11, 3227. Copyright 2011 American Chemical Society.)

2. We assume that the graphene sheet adheres perfectly to substrate and that as temperature is increased it is therefore forced to expand, with a rate of expansion exactly determined by the known thermal expansion coefficient of the substrate. But in reality the substrates used usually consist of layers of different materials and it is not clear which layer will determine the thermal expansion of the graphene. For instance, in the case of an Si/SiO_2 substrate, the 300 nm thick layer of SiO_2 (on which the graphene lies) is bonded to a much thicker layer of Si, typically ~500 μm thick. Some studies have assumed that the expansion/contraction of the graphene layer as temperature or pressure is varied is determined by the characteristics of the SiO_2 layer [8] whilst others have assumed it is determined by the characteristics of the underlying Si ([20] of Chapter 9).

3. We understand the negative thermal expansion coefficient to be a consequence of the presence of out-of-plane phonons which would exist in free-standing graphene, but it is not known to what extent these are suppressed by the presence of a substrate.

4. To calculate the graphene G peak biaxial strain coefficient, we use a value for the Grüneisen parameter [8] obtained experimentally. There is a significant disagreement in the literature regarding the value of the Grüneisen parameter (see Table 10.2).

5. Since the Grüneisen parameter has been measured only at ambient temperature, we neglect any possible variation with temperature. In other materials, the Grüneisen parameters generally do exhibit variation with temperature ([5] of Chapter 1).

In Figure 10.4, we plot the results of various theoretical and experimental works to determine the thermal expansion coefficient, including Reference 4, the findings using the data in Figure 10.3. We see massive disagreement between the various works but a consensus that, at least up to ambient temperature, the thermal expansion coefficient is negative. Above ambient temperature, it is likely that the thermal expansion coefficient becomes positive [6].

10.1.3   Graphene thermal expansion coefficient: Comparison to graphite

Given the difficulty in making an accurate measurement of the thermal expansion coefficient of graphene, it is worthwhile to make a comparison to 3D graphite. In graphite, after all, we can make an accurate measurement of all lattice constants in a free-standing sample at any temperature up to the sublimation point using X-ray diffraction.

Earlier, we concluded that a 3D material usually has a positive thermal expansion coefficient defined in terms of the change of volume as a function of temperature (Equation 10.1). But a sample can contract upon heating along one axis, or even in one plane, and the expansion in the other axis/plane can be sufficient to ensure that the thermal expansion coefficient defined in terms of volume is still positive. Thus, negative biaxial and uniaxial [9] thermal expansion coefficients have frequently been observed in 3D materials. In graphite, we would expect the thermal expansion behaviour in the xy plane to be different to that in the z-axis, due to the very different nature of the phonons excited in the in-plane and out-of-plane directions. It turns out that the thermal expansion coefficient of graphite within the xy plane is also negative at low temperature, according to both experiment and first principles calculations (Figure 10.5).

10.1.4   Effects of negative thermal expansion in graphene

Whilst it is not possible to make an accurate quantitative determination of the thermal expansion coefficient of graphene, in a qualitative manner the effects of the negative thermal expansion are observed in a wide variety of experiments.

For instance, in the last chapter, we discussed the growth of graphene on copper by chemical vapour deposition.

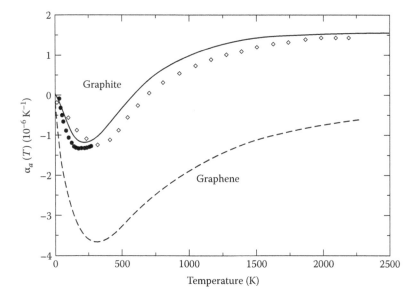

FIGURE 10.5
Calculated in-plane thermal expansion coefficients of graphite (solid line) and graphene (dotted line). Predicted negative thermal expansion coefficient for graphite is in reasonable agreement with available experimental results (black and white diamonds). (Reprinted with permission from Mounet, N. and Marzari, N., *Phys. Rev. B*, 71, 205214. Copyright 2005 by the American Physical Society.)

The Raman spectra of graphene grown on copper using this method are essentially identical to those of graphene produced using the mechanical exfoliation technique. This indicates that the graphene is not under significant strain due to the substrate. However, when graphene is grown instead on a silicon carbide substrate by epitaxial growth ([18] of Chapter 9), it appears that the interaction between the graphene and the substrate is much stronger.

The growth on silicon carbide takes place at a temperature of 1250°C–2000°C. Following growth, the SiC substrate contracts upon cooling and forces the graphene to contract also, even though the graphene would expand upon cooling if it were free-standing. The evidence for this is in the Raman spectra of graphene samples grown by this method; both the G and 2D peaks are shifted to higher wavenumbers (7–10 cm⁻¹). To give some idea of the in-plane strain this corresponds to, it requires several gigapascals of hydrostatic pressure to induce this shift in the Raman peaks of bulk free-standing graphite ([12] of Chapter 7).

10.1.5 Thermal expansion and phonon frequencies under strain

On a macroscopic scale, the effects that lead to graphite having a negative thermal expansion coefficient can be illustrated using an elastic band. Let us consider the frequency of the low-energy out-of-plane acoustic phonons thermally excited in graphene at low temperature and ambient temperature which determine the thermal expansion behaviour under these conditions. Analogous to Equations 10.8 and 10.9, we can relate the shift in acoustic phonon frequency $\Delta\omega_{Ac}$ to the strain ε in the graphene lattice and phonon frequency at zero strain ω_{Ac}:

$$\Delta\omega_{Ac} = \beta_{Ac}\varepsilon \qquad (10.10)$$

where $\beta_{Ac} = -2\omega_{Ac}\gamma_{Ac}$, the strain coefficient for this phonon. Except that, in this case, the Grüneisen parameter γ_{Ac} is negative. This is a direct consequence of the negative thermal expansion coefficient of graphene; the thermal expansion coefficient of a material is directly proportional to the weighted average of the Grüneisen parameters of the phonons excited at the sample temperature ([5] of Chapter 1). Hence the negative thermal expansion coefficient of graphene necessitates a negative Grüneisen parameter for the low-frequency out-of-plane acoustic phonons and therefore that the phonon frequency increases under tensile strain (positive ε in Equation 10.10).

An increase in phonon frequency under tensile strain is in fact highly unusual behaviour; in 3D materials phonon frequencies generally decrease under tensile strain and increase under compressive strain as a consequence of the positive Grüneisen parameter. However, it is the behaviour which we observe in a common 1D system on a macroscopic level. When we stretch an elastic band, it vibrates at a higher frequency! Readers may wish to refer to the bibliography for further reading in the standard solid state physics textbooks about the Grüneisen parameter.

10.2   Thermodynamic stability of graphene

The effect of the low-frequency out-of-plane phonons in graphene is not limited to causing a negative thermal expansion effect. Their presence also ensures that a completely flat free-standing sheet of graphene is thermodynamically unstable at $T > 0$ K. Such a sheet would spontaneously fold up on itself. This can be demonstrated using a simple calculation ([1] of Chapter 3) to predict the number of such phonons that will be present in a graphene sheet at finite temperature. We begin with Equation 10.6 (reproduced in the following). This gives, for phonons along a given dispersion relation $E(k)$, the number N of phonons excited at temperature T. This is determined by the Bose–Einstein distribution (first term) multiplied by the phonon density of states (second term). We integrate over all possible values of k; from the minimum value k_{min} which is determined by the maximum wavelength phonon that will fit onto the finite graphene sheet, up to infinity. The inclusion of the energy dispersion relation $E(k)$ in the Bose–Einstein distribution ensures that the lack of thermal energy available to excite the higher energy phonons is accounted for.

$$N = \int_{k_{min}}^{\infty} \frac{1}{e^{\frac{E(k)}{k_BT}} - 1} \times \frac{k}{2\pi} dk$$

We begin by including the specific form of the dispersion relation for these phonons (Figure 10.2). We discussed in Chapter 6 how, as we move away from the Γ point, E is directly proportional to k^2 in this branch of the dispersion relation. So we can write

$$E = \alpha k^2$$

where α is a constant. We can also write that $\beta = (1/k_BT)$ to obtain

$$N = \frac{1}{2\pi} \int_{k_{min}}^{\infty} \frac{kdk}{e^{\alpha\beta k^2} - 1}$$

Making the substitution $x = k^2$, we obtain

$$N = \frac{1}{4\pi} \int_{k_{min}^2}^{\infty} \frac{dx}{e^{\alpha\beta x} - 1}$$

This is a standard integral that can be looked up in the literature,* giving

$$N = \frac{1}{4\pi} \left[-\frac{\alpha\beta x - \ln\left[e^{\alpha\beta x} - 1\right]}{\alpha\beta} \right]_{k_{min}^2}^{\infty} \qquad (10.11)$$

At the $x = \infty$ limit, the numerator in Equation 10.11 is zero so we need to only evaluate the lower limit. We therefore obtain

$$N = \frac{1}{4\pi} \times \frac{1}{\alpha\beta} \left[\alpha\beta k_{min}^2 - \ln\left[e^{\alpha\beta k_{min}^2} - 1\right] \right] \qquad (10.12)$$

We now define a new variable L_T, such that

$$\frac{L_T^2}{L^2} = \alpha\beta k_{min}^2$$

Recalling that k_{min} is determined by the length L of the sheet ($k_{min} = 2\pi/L$), we find that

$$k_BT \approx \alpha \times \left(\frac{2\pi}{L_T}\right)^2$$

---

* $\int \frac{dx}{a + be^{mx}} = \frac{mx - \ln(a + be^{mx})}{am}$ from Reference [56].

Hence L_T is the wavelength of the highest energy phonons which can be excited at temperature T. Hence, the physical significance of L_T is that it is the wavelength of the highest energy phonons that can be excited in significant quantity at temperature T.

We can then substitute into Equation 10.12 to obtain

$$N \approx \frac{1}{4\pi} \times \left[k_{min}^2 - \frac{k_B T}{\alpha} \ln \left[e^{\frac{L_T^2}{L^2}} - 1 \right] \right]$$

On an atomic scale, real graphene sheets are very large (even using the mechanical exfoliation technique, up to ~10 μm diameter). Hence,

$$\frac{L_T^2}{L^2} \ll 1$$

We can perform a binomial expansion on the exponential term in the previous equation to obtain

$$N \approx \frac{1}{4\pi} \times \left[k_{min}^2 - \frac{k_B T}{\alpha} \ln \left[\frac{L_T^2}{L^2} \right] \right]$$

In the case of L becoming very large and hence k_{min} becoming very small, the second term in this equation dominates and we can obtain our final expression for the number of phonons in the graphene sheet:

$$N \approx \frac{1}{2\pi} \times \frac{k_B T}{\alpha} \ln \left[\frac{L}{L_T} \right] \tag{10.13}$$

We have already established that $L \gg L_T$ and hence $N \to +\infty$ whenever $T > 0$ K. Thus at any finite temperature an infinite number of out-of-plane phonons should appear in a free-standing graphene sheet of the size typically synthesized or exfoliated in the laboratory. The key unusual features of the original expression for N (Equation 10.6) that led to this result were the form of the dispersion relation for the phonons in question ($E \propto k^2$) and the nature of the density of states in a 2D material (Equations 10.6 and 3.19) which exhibits a dependence on k different to that in a 3D material.

So how is a real graphene sheet stabilized when an infinite number of out-of-plane phonons should be present? It is through a combination of the following three factors:

1. The presence of microscopic ripples to disrupt the perfect 2D nature of the graphene sheet. The presence of these ripples has been confirmed with a variety of experimental techniques (for instance, electron

diffraction and scanning tunnelling microscopy as discussed in Chapter 8).

2. The graphene sheet bonding to a substrate. The bonding of the graphene sheet to a substrate via the van der Waals interaction also plays a role in ensuring that the sheet does not spontaneously fold up on itself. This is the case even if there is an aperture in the substrate (Chapter 9).

3. Suspension of the graphene flake in a liquid. Our concept from everyday life that a liquid is not viscous and allows objects to pass through it with ease is only valid on a macroscopic scale. If you were just 10 nm tall and you were swimming through water, it would feel like swimming through treacle. On a microscopic scale, typical liquids are very viscous and this viscosity can prevent a mono-layer graphene flake suspended in the liquid from folding.

10.2.1 Comparison to SWCNTs

How is the behaviour of this out-of-plane phonon mode different in the case of SWCNTs? In a graphene sheet, what is happening physically when for the out-of-plane acoustic phonon discussed earlier, $k \to 0$ (i.e. we move towards the Γ point)? This corresponds to the phonon wavelength in the plane of the graphene sheet becoming very large and the energy becoming very small.

The equivalent of this vibration in a carbon nanotube is the radial breathing mode, and, as discussed extensively in Chapters 6 and 7, this vibration has non-zero frequency and non-zero energy at the Γ point. It corresponds to the entire tube expanding and then contracting. Therefore, we cannot excite an infinite number of these phonons at $T > 0$ K and those phonons which are excited do not destroy the nanotube.

For these reasons a free-standing SWCNT is stable, unlike free-standing graphene.

10.3 Mechanical properties of graphene

10.3.1 Mechanical properties of graphite

It is worthwhile to begin our discussion of the mechanical properties of graphene with reference to the mechanical properties of graphite due to the much wider range of experimental measurements that are possible on graphite compared to graphene. In particular, it is possible to accurately measure the compressibility (bulk modulus) of free-standing graphite in the in-plane and out-of-plane directions.

This can be done by performing angle-dispersive synchro-tron X-ray diffraction on graphite at high quasi-hydrostatic pressure inside a diamond anvil high pressure cell ([12] of Chapter 7 and [10]). Both the measurement of the graphite lattice constants using X-ray diffraction and the measurement of pressure using the photoluminescence from a ruby crystal located close to the graphite in the sample chamber are highly accurate.

Figure 10.6 shows the lattice constants of graphite in the in-plane and out-of-plane direction as a function of quasi-hydrostatic pressure, measured using X-ray diffraction ([12] of Chapter 7).

As expected, we can see that graphite is highly com-pressible in the out-of-plane direction and somewhat less compressible in the in-plane direction. To formalize this observation, we can fit the lattice constants $a(P)$ and $c(P)$ sepa-rately to 1D Murnaghan equations of state [11,12]:

$$\frac{r(P)}{r_0} = \left[1 + \frac{\beta'}{\beta_0} P\right]^{-1/\beta'} \tag{10.14}$$

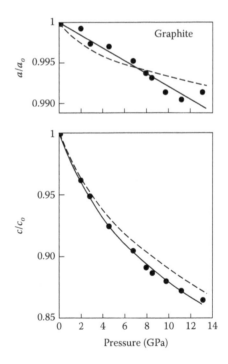

FIGURE 10.6

Lattice constants of graphite in the in-plane direction (a, upper panel) and out-of-plane direction (c, lower panel) as a function of hydrostatic pressure. (Reprinted with permission from Hanfland, M., Beister, H. and Syassen, K., *Phys. Rev. B*, 39, 12598 [Chapter 7]. Copyright 1989 by the American Physical Society.)

TABLE 10.1

Bulk modulus β_0 and pressure derivative β' of graphite along the a- and c-axis

	β_0 (GPa^{-1})	β' (Dimensionless)
a-axis	1250 ± 70	1 (fixed)
c-axis	35.7 ± 0.3	10.8 ± 0.9

Source: Hanfland, M. et al., *Phys. Rev. B*, 39, 12598, 1989.

Physically, β_0 represents the bulk modulus along the relevant axis and β' is the pressure derivative of β_0. The experimentally obtained values of β_0 and β' along the a-axis and c-axis are tabulated in Table 10.1.

Along the c-axis, one can fit a Murnaghan equation of state (as shown in Figure 10.6 lower panel) to obtain values for β_0 and β'. Along the a-axis, however, the graphite is so incompressible that it is not possible to fit an equation of state with two adjustable parameters. The change in lattice constant is small and cannot be measured with adequate precision. Instead β' is fixed and we treat the compressibility as constant. Despite the relatively large experimental error, we can observe that the bulk modulus for compression of graphite in the in-plane direction rivals that of diamond.

The Raman spectra of graphite as a function of pressure have also been measured ([12] of Chapter 7 and [13]). Both the Raman-active phonon modes in pristine graphite (the G peak and the 2D peak) correspond to in-plane vibrations. The pressure-induced shifts in the Raman peaks are therefore interpreted as being due to the in-plane strain ε in the graphene sheet:

$$\frac{\omega_i(\varepsilon)}{\omega_i(\varepsilon = 0)} = \left[\frac{a(\varepsilon)}{a(\varepsilon = 0)}\right]^{-2\gamma_i} \tag{10.15}$$

where the Grüneisen parameter of the ith phonon mode γ_i is the dimensionless parameter linking the strain in the graphene sheet to the Raman peak position under strain $\omega_i(\varepsilon)$. The Grüneisen parameters for the E_{2g} mode (G peak) and A_{1g} mode (2D peak) can therefore be calculated. We obtain $\gamma_G = 1.59$ ([12] of Chapter 7 and [20] of Chapter 9) and $\gamma_{2D} = 2.84$ ([12] of Chapter 7, [20] of Chapter 9 and [13]).

It is important to understand the limitations of these measurements. In the case of γ_G, we do make a direct measurement of the Grüneisen parameter for a specific phonon at the centre of the Brillouin zone.

In the case of γ_{2D}, however, this is not the case. The double-resonant Raman scattering process that produces the 2D peak involves the excitation of an electron from the valence band to the conduction band and the creation of two phonons that are not at the centre of the Brillouin zone as discussed at length in Chapter 7. When we apply pressure to the graphite crystal, we expect the electronic dispersion relation to change slightly; whilst the energy of the laser photon used in the process remains constant. Hence, the conservation of energy and momentum dictates that, as we increase pressure, the double-resonant Raman scattering process that produces the 2D peak will occur for a different phonon.

Hence the measured value of γ_{2D} is not a characteristic of a specific phonon, as would generally be the case for a Grüneisen parameter.

Before moving onto graphene, it is appropriate to make one final remark regarding graphite at high pressure. Historically, carbon has perhaps been compressed more than any other element due to the possibility of using high pressure to transform graphite (or any other carbon-rich material for that matter) into diamond [14,15]. The bulk graphite $\rightarrow$ diamond phase transition requires a temperature in excess of 1500 K to ensure there is enough thermal energy available for the transition to take place. Hence when we compress graphite at ambient temperature, we do not see a transition to the diamond phase despite the fact that a pressure of just ≈ 5 GPa is required to enter the region of the phase diagram where diamond is the thermodynamically stable phase of carbon (see Section 10.5).

So when graphite is compressed at ambient temperature, there is not adequate thermal energy available to convert to the diamond phase, but the graphite phase becomes increasingly unstable. The result of these conditions is an increasing level of structural disorder as pressure is increased; some authors have described graphite compressed above 20 GPa at ambient temperature as amorphous [13]. The consequence of the increasing level of structural disorder is a large drop in the intensity of both X-ray diffraction and Raman peaks as pressure is increased.

10.3.2   Graphene under strain

The study of the mechanical properties of a single graphene layer is hindered by the fact that a single layer of free-standing graphene is thermodynamically unstable and also by the fact that an X-ray diffraction experiment on a single layer of graphene is not possible. Therefore, we cannot perform an equivalent experiment on a single layer of graphene to the high pressure experiments on graphite described in the previous section.

On the other hand, the fact that graphene is the thinnest material known opens up possibilities for a very different range of experiments to probe its mechanical properties. Due to the exceptional surface-area-to-volume ratio of graphene, we can apply massive stresses (both tensile and compressive) to the single atomic layer simply by placing it on

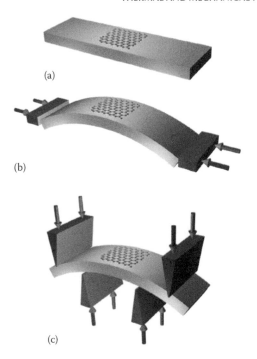

FIGURE 10.7
Schematic diagram of a uniaxial strain experiment on mono-layer graphene. The graphene is deposited onto a flexible substrate (a) before strain is applied to the graphene sample by bending the substrate (b or c). (Reprinted with permission from Mohiuddin, T.M.G. et al., *Phys. Rev. B*, 79, 205433 [Chapter 7]. Copyright 2009 by the American Physical Society.)

a suitable substrate and compressing or stretching the substrate (Figure 10.7). Raman spectroscopy can then be performed on graphene *in situ* under strain. The strain-induced shift and, in some cases, splitting of the Raman peaks can be observed. Tensile and compressive strain can be applied to graphene in both uniaxial and biaxial arrangements. Lastly, graphene on a substrate can be subjected to hydrostatic pressure in the diamond anvil high pressure cell ([20] of Chapter 9 and [16,17]).

Since we do not measure the compressibility of the graphene sample when we stretch or compress graphene on a substrate, the main value of such experiments is in the study of the phonons in graphene as a function of strain; measurement of the Grüneisen parameters using uniaxial and biaxial strain and breaking of symmetry in the phonon dispersion relation in the case of uniaxial strain. These parameters are of value as they provide a means to measure strain in graphene – we have seen in Section 10.1 how the measurement of strain in graphene is crucial to determine the thermal expansion coefficient. The measurement of strain in graphene is also important in the realization of potential electronics applications [18], the development of graphene-based composite materials [19] and the use of graphene itself as a strain sensor.

Table 10.2 lists the various values obtained for the Grüneisen parameters of the Raman-active phonon modes in graphene with some notes about the experimental/theoretical conditions for the different works.

As is evident from Table 10.2, a wide variety of values for the Grüneisen parameters have been proposed. We expect to obtain more reliable values for the parameter γ_G since the value of this parameter is determined only by mechanical considerations, while the measured value of γ_{2D} is affected by the change in the electronic dispersion relation under strain (just as for graphite). But the measured values of γ_G range from 0.67 [20] to 2.2 [8]. Why this variation? The list of factors that could cause uncertainty in these experimental measurements is long. For instance, we assume in all cases that the graphene sample remains adhered to the substrate while strain is applied, but do not verify this directly. In the case of an experiment under uniaxial strain, we can assume that stretching (contraction) along the axis of the applied strain will result in contraction (stretching) in the axis perpendicular to the applied strain. The ratio of these strains is given by the Poisson's ratio. But should we use the Poisson's ratio of graphene or of the substrate?

TABLE 10.2

Contrasting values of the Grüneisen parameters for the graphene G, D and 2D Raman peaks (γ_G, γ_D, γ_{2D}) obtained in different experimental and theoretical works on both graphene and graphite

Study	γ_G	γ_D	γ_{2D}	Notes
Mohiuddin et al. ([6] of Chapter 7)	1.99	3.55	2.84	Uniaxial strain experiment (graphene)
Mohiuddin et al. ([6] of Chapter 7)	1.87	2.7	2.7	Density functional theory (DFT) calculation
Ding et al. [53]		2.30	2.98	Biaxial strain experiment (graphene)
Yoon et al. [8]	2.2			Uniaxial strain experiment (graphene)
Filintoglou et al. [16]	2.1			Hydrostatic pressure experiment (graphene)
Huang et al. [20]	0.67			Uniaxial strain experiment (graphene)
Ghandour et al. [35]	1.34			Uniaxial strain experiment (graphene) and hydrostatic pressure experiment (SWCNT)
Thomsen et al. [54]	2.0			DFT calculation
Goncharov [13] and Hanfland et al. ([12] of Chapter 7)			2.84	Hydrostatic pressure experiment (graphite)
Hanfland et al. ([12] of Chapter 7)	1.59			Hydrostatic pressure experiment (graphite)

In the case of hydrostatic pressure experiments, we consider the strain in the graphene-substrate system. We take the simple approach of assuming that the strain along each axis is determined only by the stress along that axis. It has, however, been suggested that a more sophisticated approach is needed in which the stress σ and strain ε are related by a tensor quantity S [17]. Hence $\sigma = S\varepsilon$. The out-of-plane stress does, according to this treatment, affect the in-plane strain.

The difficulty in accurately determining the Grüneisen parameters prevents accurate measurement of strain in graphene, in turn preventing accurate measurement of the thermal expansion coefficient and inhibiting the realization of potential applications of graphene as a strain sensor.

10.3.3   Indentation testing on graphene

In the works described in the previous section the strains applied to the graphene lattice (both tensile and compressive) are relatively small – about 1%. It is not possible to apply a significantly larger compressive strain without slippage of the sample over the substrate (in the case of graphene) or amorphization (in the case of graphite).

However, it is possible to apply much larger tensile strain to graphene. This can be achieved by preparing a mono-layer graphene sample on a substrate containing an aperture and pushing down on the graphene sample over the aperture with an atomic force microscope (AFM) tip. Using this method, it has been demonstrated that graphene can withstand a tensile strain of up to 20% (Figure 10.8) ([34] of Chapter 1).

In the experiments described, the stress and strain applied to the graphene sheet are both determined directly by measuring the load on the AFM tip, and the position of the tip as the load is increased. Thus the graph of stress versus strain shown in the lower panel of Figure 10.8 is obtained. The stress (σ) and strain (ε) can be described using a quadratic fit:

$$\sigma = E\varepsilon + D\varepsilon^2$$

For low stress the behaviour is approximately linear and can be described using the parameter E, the Young's modulus of graphene. Experimentally this is determined to be $E = 340 \pm 50$ Nm^{-1}, with $D = -690$ Nm^{-1}.

In addition, the intrinsic strength (breaking stress) of the graphene sheet is extremely large. From the data shown in Figure 10.8, the breaking stress is shown as 130 GPa ([34] of Chapter 1), establishing graphene as the strongest material ever tested in this manner. The extremely high breaking stress is a reflection of the intrinsic strength of the sp^2 C–C bond in graphene and also of the fact that it is possible to prepare a graphene sample with virtually no defects, in which the intrinsic strength of the pristine C–C bond is the only factor limiting the stress in which the sample can withstand.

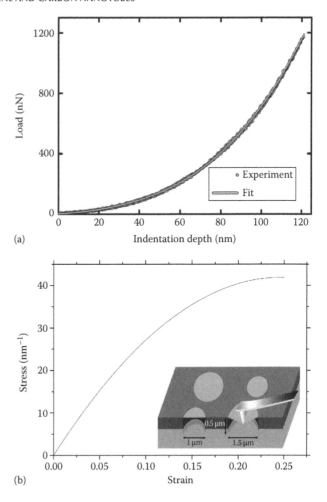

FIGURE 10.8

(a) Experimental data showing applied load versus indentation depth as graphene is indented using an atomic force microscope tip. (b) Graph of stress versus strain obtained by fitting to data from upper panel. (Inset to b) Schematic diagram of nanoindentation experiment on graphene. ([a] and inset to [b]: From Lee, C., Wei, X., Kysar, J.W.; Hone, J., *Science*, 321, 5887, 2008 [Chapter 1]. Reprinted with permission of AAAS.)

10.4 Mechanical properties of SWCNTs

10.4.1 Mechanical probing of bundles of SWCNTs

A variety of experiments have been performed in which the mechanical properties of carbon nanotubes have been measured. Initially, experiments to probe the properties of individual SWCNTs proved extremely challenging and a number of studies instead probed the properties of multi-walled carbon nanotubes [21] or bundles of SWCNTs [22].

Similar to the case of graphene discussed earlier, bundles of SWCNTs have been deposited on substrates containing

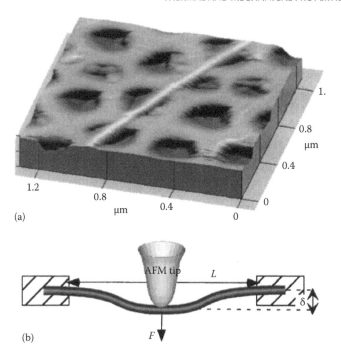

FIGURE 10.9
(a) Atomic force microscope (AFM) image of SWCNT bundle deposited on substrate containing apertures. (b) Schematic diagram of process in which strain is applied to the SWCNT bundle using the AFM tip. (Reprinted with permission from Salvetat, J.-P. et al., *Phys. Rev. Lett.*, 82, 944. Copyright 1999 by the American Physical Society.)

an aperture and probed using an AFM tip [23] (Figure 10.9). These experiments established the Young's modulus of bundles of SWCNTs to be of the order of 1 TPa. In contrast, transition metals tend to exhibit Young's moduli in the range 100–400 GPa.

A very large number of studies have focussed on the behaviour of bundled SWCNTs under quasi-hydrostatic pressure in the diamond anvil high pressure cell. A multitude of Raman spectroscopy studies (for instance, References 24–27), X-ray diffraction studies [28–30], theoretical calculations and molecular dynamics simulations (for instance, References [21] of Chapter 1 and [27,31]) have demonstrated that SWCNTs mechanically deform under hydrostatic pressure in various ways. Two reversible deformation processes are proposed at moderate pressures (below 10 GPa), and evidence has been found for irreversible changes to the SWCNT material at higher pressures (10 GPa+). Nonetheless, the complete reversibility of changes to the SWCNT material at pressures of up to 10 GPa (100 kbar) is an incredible demonstration of the resilience of the nanotube structure.

At modest high pressures, a distortion in the SWCNT cross section from a circular to a hexagonal shape (conforming to the bundle structure) has been proposed. Molecular dynamics simulations suggest that this occurs even at ambient pressure for larger-diameter tubes ([21] of Chapter 1) (Figure 10.10, inset). The deformation is expected to occur at

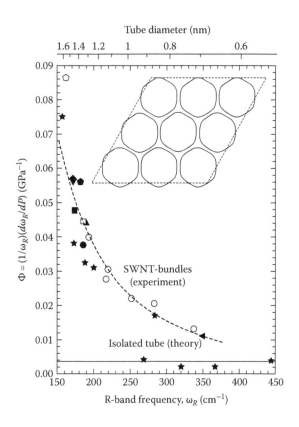

FIGURE 10.10

Normalized pressure derivative of the SWCNT R band (RBM) frequency ω_R. Open square and pentagon – molecular dynamics and force constant calculations for (9,9) and (10,10) bundled SWCNTs. Stars – experimental results for the R band of the inner nanotube in double-walled carbon nanotubes. Other data points are experimental results for different diameter bundled SWCNTs. The dotted line is a fit to the experimental data and the horizontal solid line is the theoretical prediction for isolated SWCNTs [51]. Inset: MD simulations of (30,30) SWCNT bundles at atmospheric pressure, showing polygonized cross sections. (From Venkateswaran, U.D.: *Phys. Stat. Sol. (b)*. 2004. 241. 3345. Copyright Wiley-VCH Verlag GmbH & Co. KGaA. Reproduced with permission; [Inset]: Reprinted with permission from Elliott, J.A., Sandler, J.K.W., Windle, A.H., Young, R.J. and Shaffer, M.S.P., *Phys. Rev. Lett.*, 92, 095501 [Chapter 1]. Copyright 2004 by the American Physical Society.)

ambient conditions when the energy of the increased attractive van der Waals interaction between the distorted tubes exceeds the strain energy due to the bending of the nanotube at the vertices of the hexagons.

The molecular dynamics simulations suggest that this hexagonal deformation gradually becomes more pronounced at higher pressures. This is supported by several Raman studies of the RBMs in bundled SWCNTs [26] (Figure 10.10). The pressure dependence of the RBM frequency gradually diverges from what would be expected for an isolated tube (i.e. the case where there is no polygonization) for larger tubes.

In addition to the gradual polygonization, a reversible collapse of the SWCNTs to an oval cross section is predicted by several authors in the range of 2–6 GPa. This collapse is predicted to occur at a pressure inversely proportional to the SWCNT diameter (Figure 10.11) ([21] of Chapter 1).

Experimentally, collapse of the SWCNT has been confirmed by many authors by the disappearance or discontinuous intensity change of the RBM peak at the collapse pressure, together with a reduction in the gradient of the G peaks from 7–10 cm^{-1} GPa^{-1} to a gradient closer to that observed with graphite (ca. 5 cm^{-1} GPa^{-1} – see Figure 10.12) [25,32].

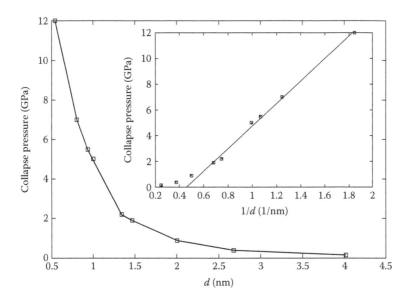

FIGURE 10.11

The prediction of molecular dynamics simulations that the collapse pressure of SWCNTs is inversely proportional to the tube diameter is shown. The data points are SWCNT diameters for which a simulation was performed. The fit in the main figure is a guide to the eye only and the fit in the inset (to reciprocal diameter) is a linear fit. (Reprinted with permission from Elliott, J.A., Sandler, J.K.W., Windle, A.H., Young, R.J. and Shaffer, M.S.P., *Phys. Rev. Lett.*, 92, 095501. Copyright 2004 by the American Physical Society.)

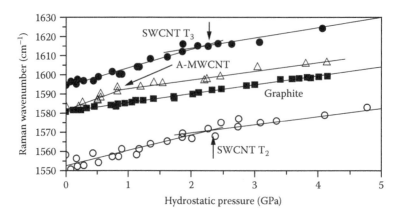

FIGURE 10.12

Graph showing pressure-induced shifts of the components of the SWCNT G peaks and equivalent modes in multi-walled carbon nanotubes and graphite. The changes in gradient labelled with the arrows for both SWCNTs and multi-walled carbon nanotubes are due to collapse of the tubes. The effect is absent in the case of graphite. (Reprinted with permission from Sandler, J. et al., *Phys. Rev. B*, 67, 035417. Copyright 2003 by the American Physical Society.)

Whilst the collapse of the SWCNT at pressure is reversible (the nanotubes re-inflate when pressure is released), at much higher pressure (ca. 13 GPa) an irreversible transition has been observed [28]. This is most likely to be due to covalent bonds forming between the adjacent nanotubes [33]. Irreversible interlinking of adjacent atomic layers in graphite has also been observed at ca. 17 GPa [10].

SWCNTs have also been studied using X-ray diffraction and Raman spectroscopy under compression in a diamond anvil high pressure cell with no pressure-transmitting medium [34]. Since SWCNTs can support large shear stresses, this leads to a substantial variation in pressure across the sample chamber and the pressure not being equal along all axes (i.e. non-hydrostatic). This is in contrast to the studies described thus far, in which the SWCNTs were compressed surrounded by a soft material such as methanol–ethanol solution which transmits quasi-hydrostatic pressure to the sample due to its inability to support large shear stresses. The study under non-hydrostatic pressure [34] found the SWCNT Raman peaks to be fully recoverable after pressurization to 30 GPa whilst the X-ray diffraction peak corresponding to the triangular SWCNT bundle lattice was irreversibly destroyed by pressurization to only 5 GPa. It was therefore concluded that the SWCNT lattice was easily destroyed by non-hydrostatic pressure whilst the individual SWCNTs were remarkably resilient to it.

It is worthwhile, however, to exercise caution when interpreting the findings of high pressure experiments on SWCNTs; especially the findings obtained using optical spectroscopy. As discussed in Chapter 7, all of the observed

Raman peaks from SWCNTs originate from a resonant Raman scattering process involving the promotion of an electron from the valence band to the conduction band. When we take the Raman spectrum of a sample of bundled SWCNTs, we are collecting the spectrum from some small proportion of the SWCNTs in the sample, those for which the double resonance condition is met. As we increase pressure in a hydrostatic pressure experiment, we expect the electronic dispersion relation to change. Hence, at different pressures but constant laser excitation energy the double resonance condition is met for different phonons in different tubes [35]. This considerably complicates the interpretation of Raman data on SWCNTs at high pressure.

As if that was not enough of a problem, a (real) bundle of SWCNTs contains tubes with a certain distribution of diameters (Chapter 9), so its behaviour under pressure may be quite complex and not modelled accurately by a theoretical calculation performed on a bundle of SWCNTs all having the same (n,m). Any reader who still believes that the system can be well understood may wish to consider the effect of interaction between the nanotubes and the pressure-transmitting medium on the experimental observations ([13] of Chapter 5 and [36–38]).

10.4.2 Mechanical probing of individual SWCNTs

Due to the 1D nature of SWCNTs, it has proved possible to utilize some novel methods to probe the mechanical properties of individual tubes. For instance, the thermal vibration of individual SWCNTs with one end tethered and one end free to vibrate have been studied to measure the Young's modulus of an individual SWCNT along the tube axis [39,40], with values of the order of 1 TPa obtained (albeit with a large variation in the measured values). Figure 10.13 shows examples of the vibrations observed in these studies and a histogram of measured values.

When force is exerted perpendicular to the tube axis, on the other hand, SWCNTs have been demonstrated to be far softer. Experiments in which force is applied to a single SWCNT in this direction using a scanning tunnelling microscope tip have revealed the Young's modulus along this axis to be an order of magnitude lower than that in the axial direction [41].

Techniques have also been developed to probe individual SWCNTs under hydrostatic pressure. Essentially, there are two methods. It is possible to separate SWCNTs from the bundle as described in Chapter 9 and place isolated SWCNTs into the high pressure cell in solution surrounded by surfactant to prevent reaggregation into the bundle ([13] of Chapter 5 and [42]). On the other hand, it is possible to pressurize bundled SWCNTs and collect the Raman spectra from tubes with the same (n,m) as pressure is increased. This is achieved by

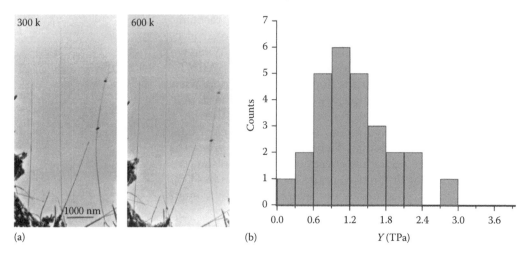

FIGURE 10.13

(a) Bright field transmission electron microscope (TEM) image of thermal vibrations of a SWCNT at ambient temperature and at 600 K. The blurring in the image is due to the thermal vibration, which increases at 600 K. (b) Histogram of measured Young's modulus of different SWCNTs studied in a similar experiment. ([a]: Reprinted by permission from Macmillan Publishers Ltd. *Nature,* Treacy, M.M.J., Ebbesen, T.W. and Gibson, J.M., 381, 678, copyright 1996; [b]: Reprinted with permission from Krishnan, A., Dujardin, E., Ebbesen, T.W., Yianilos, P.N. and Treacy, M.M.J., *Phys. Rev. B,* 58, 14013. Copyright 1998 by the American Physical Society.)

utilizing a tunable laser to change the incident photon energy in line with the changes in the electronic dispersion relation as pressure is increased [35].

Scientifically, this work has provided a direct confirmation of the expected correlation between SWCNT diameter and collapse pressure [42] since individual tubes can be tracked as pressure is increased. In addition, it has provided insights into the role of excitons in the photoluminescence observed from isolated nanotubes. To account for the shift in observed photoluminescence energies as a function of pressure, it was necessary to account for the effect of the SWCNT's environment on the excitons responsible for the observation of the photoluminescence. The field lines of the electron–hole pair comprising the exciton extend outside the SWCNT and thus the exciton energy is influenced by the permittivity of the environment of the SWCNT. Under high pressure the change in permittivity of the pressure medium surrounding the SWCNTs was observed to affect the exciton energies ([13] of Chapter 5).

10.4.3 Stability of large SWCNTs

The study of SWCNTs at high pressure provides insights into the stability of large SWCNTs at ambient conditions. Figure 10.11 showed the theoretically expected correlation between collapse pressure and tube diameter, which has been

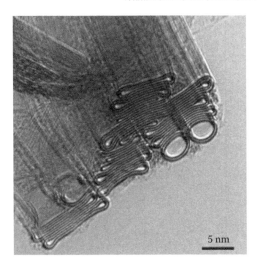

5 nm

FIGURE 10.14
TEM image of mainly double-walled carbon nanotubes
under vacuum demonstrating collapse of the larger tubes
(left and centre) whilst smaller tubes remain inflated (right).
(From Motta, M., Moisala, A., Kinloch, I.A. and Windle,
A.H.: *Adv. Mater.* 2007. 19. 3721. Copyright Wiley-VCH Verlag
GmbH & Co. KGaA. Reproduced with permission.)

verified experimentally in several works. This is in agreement
with the experimental observation at ambient conditions
([20,21] of Chapter 1) that the largest inflated (i.e. cylindrical)
SWCNT observed experimentally is approximately 2.5 nm
in diameter. For double-walled carbon nanotubes also, it is
observed that there is a maximum diameter of tube which is
stable at ambient conditions. This is illustrated very nicely in
the transmission electron microscope (TEM) image in Figure
10.14 showing smaller double-walled nanotubes inflated
whilst the larger ones are collapsed.

10.5 High-pressure behaviour of graphene, SWCNTs and fullerenes in the context of the carbon phase diagram

To understand the behaviour of graphene, SWCNTs and fullerenes at high pressure, it is frequently useful to consider them
in the context of the pressure–temperature (P–T) phase diagram of bulk carbon. At experimentally achievable pressures
and temperatures, the thermodynamically stable phases of
bulk carbon are believed to be graphite and diamond [14,15,43]
(Figure 10.15). Note that graphite is the thermodynamically stable phase of bulk carbon at ambient conditions. So technically
speaking, diamond is not forever. Arguably it would be more
romantic to purchase a graphite ring for your loved one.

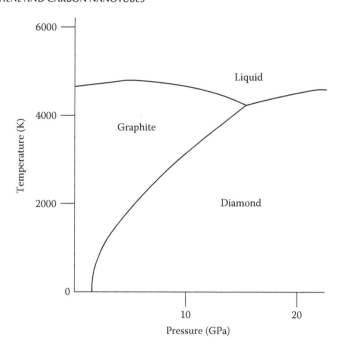

FIGURE 10.15
Schematic diagram of the pressure–temperature phase diagram of bulk carbon.

On the phase diagram there exist two separate P–T regions, a region where graphite is the thermodynamically stable phase and a region where diamond is the thermodynamically stable phase. It is important to be clear about what this distinction means. If we throw a large number of isolated carbon atoms together at a P–T point in the graphite region they will form graphite, if we throw a large number of isolated carbon atoms together in the diamond region they will form diamond.

But it does not follow from this that if we take a crystal of graphite and subject it to P–T conditions in the diamond region, it will transform into diamond. This is because there is a substantial kinetic barrier to the graphite ↔ diamond transition taking place. If graphite is compressed into the diamond region at ambient temperature it will not spontaneously transform into diamond. The application of high temperature (and, commercially, the use of a catalyst) is also required to overcome the kinetic barrier and cause the transition to the diamond phase.

Of course, the reverse is also true. When diamond is synthesized under high P–T conditions (either in the factory or in the earth's mantle), it does not spontaneously transform into graphite once it is returned to ambient conditions.

One can understand many features of the behaviour of graphene, SWCNTs and fullerenes under extreme conditions in terms of the carbon phase diagram and a frustrated transition from a graphite-like to a diamond-like phase. At ambient conditions, both SWCNTs and fullerenes exhibit a mixture of sp^2 and sp^3 hybridization. Whilst graphite can be compressed to 20 GPa at ambient temperature without exhibiting a transition to a diamond-like phase C_{60}, on the other hand, does transform to (albeit poor quality) diamond at 20 GPa at ambient temperature [44]. This can be understood as a consequence of the fact that C_{60} already exhibits some component of sp^3 hybridization thus lowering the kinetic barrier to the transition to the diamond phase.

SWCNTs exhibit a lower proportion of sp^3 hybridization than fullerenes due to the lower curvature. When pressurized under ambient temperature, SWCNTs collapse whilst still in/near the graphite region of the phase diagram, which is likely to reduce further the sp^3 component of the bonds. Hence, SWCNTs do not exhibit the transition to diamond at ambient temperature observed in fullerenes. The interlinking of the SWCNTs observed in high pressure experiments however (see earlier discussion) can be interpreted as a kinetically inhibited transition to diamond.

If the bulk graphite $\rightarrow$ diamond phase transition at high pressure is frustrated by a large kinetic barrier, what about the transition from few-layer graphene to nanodiamond? Theoretically, we expect this transition to be subject to an even larger kinetic barrier than that for bulk carbon due to the influence of surface energy – the energy cost due to broken bonds on the surface of the sample [45–48].

A crystal of few-layer graphene has extremely small surface energy. The π-bonds cost little energy to break and the only broken σ-bonds are the relatively small number of bonds at the edges of the graphene sheet. However, if this crystal was transformed into nanodiamond the sample would then have a much larger surface energy, because all the broken bonds on the surface of the sample would be strong σ-bonds [47].

On the other hand, hydrogenated graphene materials can be understood as a one-atom thick slice through the diamond lattice in which the broken σ-bonds are terminated with hydrogen [47]. A recent study [49] exploited this to synthesize hydrogenated graphene using high pressure and temperature by compressing and heating graphene in a hydrogen atmosphere; overcoming the energetic barrier against a transformation to a diamond-like phase using the hydrogen to terminate the broken σ-bonds on the surface of the sample.

References

1. W Miller, CW Smith, DS Mackenzie and KE Evans, *J. Mater. Sci.* **44**, 5441 (2009).

2. IM Lifshitz, *Zh. Eksp. Teor. Fiz. (JETP)* **22**, 475 (1952).

3. M Pozzo, D Alfè, P Lacovig, P Hofmann, S Lizzit and A Baraldi, *Phys. Rev. Lett.* **106**, 135501 (2011).

4. D Yoon, Y-W Son and H Cheong, *Nano Lett.* **11**, 3227 (2011).

5. W Bao et al., *Nat. Nanotechnol.* **4**, 562 (2009).

6. S Linas et al., *Phys. Rev. B* **91**, 075426 (2015).

7. N Bonini, M Lazzeri, N Marzari and F Mauri, *Phys. Rev. Lett.* **99**, 176802 (2007).

8. D Yoon, Y-W Son and H Cheong, *Phys. Rev. Lett.* **106**, 155502 (2011).

9. MS Senn et al., *Phys. Rev. Lett.* **114**, 035701 (2015).

10. WL Mao et al., *Science* **302**, 425 (2003).

11. FD Murnaghan, *PNAS* **30**, 244 (1944).

12. WB Holzapfel, Equations of state for solids over wide ranges of pressure and temperature, in *High-Pressure Physics*, J. Loveday (ed.), CRC Press , Boca Raton (2012).

13. AF Goncharov, *JETP Lett.* **51**, 418 (1990).

14. RM Hazen, *The Diamond Makers*, CUP, Cambridge (1999).

15. EYu Tonkov and EG Ponyatovsky, *Phase Transformations of the Elements under High Pressure*, CRC Press, Boca Raton (2005).

16. K Filintoglou et al., *Phys. Rev. B* **88**, 045418 (2013).

17. J Nicolle, D Machon, P Poncharal, O Pierre-Louis and A San Miguel, *Nano Lett.* **11**, 3564 (2011).

18. G Gui et al., *Phys. Rev. B* **78**, 075435 (2008).

19. RJ Young, IA Kinloch, L Gong and KS Novoselov, *Compos. Sci. Technol.* **72**, 1459 (2012).

20. MY Huang et al., *PNAS* **106**, 7304 (2009).

21. M-F Yu, T Kowalewski and RS Ruoff, *Phys. Rev. Lett.* **85**, 1456 (2000).

22. M-F Yu, BS Files, S Arepalli and RS Ruoff, *Phys. Rev. Lett.* **84**, 5552 (2000).

23. J-P Salvetat et al., *Phys. Rev. Lett.* **82**, 944 (1999).

24. PV Teredesai, AK Sood, DVS Muthu, R Sen, A Govindaraj and CNR Rao, *Chem. Phys. Lett.* **319** 296 (2000).

25. J Sandler et al., *Phys. Rev. B* **67**, 035417 (2003).

26. UD Venkateswaran, *Phys. Stat. Sol.* (*b*) **241**, 3345 (2004).

27. UD Venkateswaran et al., *Phys. Rev. B* **59**, 10928 (1999).

28. SM Sharma, S Karmakar, SK Sikka, PV Teredesai, AK Sood, A Govindaraj and CNR Rao, *Phys. Rev. B* **63**, 205417 (2001).

29. S Kawasaki et al., *J. Phys. Chem. Sol.* **65**, 327 (2004).

30. J Tang, L Qin, T Sasaki, M Yudasaka, A Matsushita and S Iijima, *Phys. Rev. Lett.* **85**, 1887 (2000).

31. M Sluiter and Y Kawazoc, *Phys. Rev. B* **69**, 224111 (2004).

32. MJ Peters, LE McNeil, JP Lu and D Kahn, *Phys. Rev. B* **61**, 5939 (2000).

33. T Yildirim, O Gülseren, C Kiliç and S Ciraci, *Phys. Rev. B* **62**, 12648 (2000).

34. S Karmakar et al., *New J. Phy.* **5**, 143 (2003).

35. A Ghandour et al., *Phys. Rev. B* **87**, 085416 (2013).

36. MS Amer, MS El-Ashry and JF Maguire, *J. Chem. Phys.* **121**, 2752 (2004).

37. JE Proctor, MP Halsall, A Ghandour and DJ Dunstan, *J. Phys. Chem. Sol.* **67**, 2468 (2006).

38. YW Sun et al., *J. Phys. Chem. C* **120**, 1863 (2016).

39. MMJ Treacy, TW Ebbesen and JM Gibson, *Nature* **381**, 678 (1996).

40. A Krishnan, E Dujardin, TW Ebbesen, PN Yianilos and MMJ Treacy, *Phys. Rev. B* **58**, 14013 (1998).

41. YH Wang and WZ Li, *Appl. Phys. Lett.* **98**, 041901 (2011).

42. S Lebedkin, K Arnold, O Kiowski, F Hennrich, and MM Kappes, *Phys. Rev. B* 73, 094109 (2006).

43. LM Ghiringhelli, JH Los, EJ Meijer, A Fasolino and D Frenkel, *Phys. Rev. Lett.* **94**, 145701 (2005).

44. MN Regueiro, P Monceau and J-L Hodeau, *Nature* **355**, 237 (1992).

45. AN Goldstein et al., *Science* **256**, 1425 (1992).

46. SH Tolbert and AP Alivisatos, *J. Chem. Phys.* **102**, 4642 (1995).

47. AG Kvashnin, LA Chernatonskii, BI Yakobson and PB Sorokin, *Nano Lett.* **14**, 676 (2014).

48. IF Crowe et al., *J. Appl. Phys.* **109**, 083534 (2011).

49. D Smith et al., *ACS Nano* **9**, 8279 (2015).

50. N Mounet and N Marzari, *Phys. Rev. B* **71**, 205214 (2005).

51. UD Venkateswaran et al., *Phys. Rev. B* **68**, 241406(R) (2003).

52. M Motta, A Moisala, IA Kinloch and AH Windle, *Adv. Mater.* **19**, 3721 (2007).

53. F Ding et al., *Nano Lett.* **10**, 3453 (2010).

54. C Thomsen, S Reich and P Ordejón, *Phys. Rev. B* **65**, 073403 (2002).

55. L Lyons, *All You Wanted to Know about Mathematics but Were Afraid to Ask (Mathematics for Science Students)*, CUP, Cambridge (1998).

56. IS Gradshteyn and IM Ryzhik, *Table of Integrals, Series and Products*, Academic Press, Cambridge, MA (2007).

Chemical Modification of Graphene

11.1 Overview

A pristine sheet of mono-layer graphene has a higher surface area-to-volume ratio than any other material existing in nature. The bonding of other atoms and molecules to graphene is therefore of fundamental scientific as well as technological interest. When discussing this topic, it is important to differentiate between the bonding of other atoms and molecules to pristine graphene and the bonding of other atoms and molecules to defective or disordered graphene. It is also important to differentiate between physisorption (atoms or molecules simply resting on the graphene surface, attached only via van der Waals bonds) and chemisorption (atoms or molecules forming strong covalent or ionic bonds to the graphene sheet).

In this chapter, we will concentrate on the chemisorption of atoms onto pristine mono-layer graphene, as it is for this system that one can obtain a clear-cut understanding that relates to the underlying physics.

A pristine sheet of defect-free mono-layer graphene is a relatively unreactive material. Whilst many atoms and molecules can be physisorbed on the graphene surface, it will form chemical bonds (i.e. ionic/covalent) only to hydrogen ([1] of Chapter 1), oxygen [1] and halogens [2]. Disordered few-layer graphene samples (particularly those produced by the reduction of graphite oxide) will, on the other hand, bond to a wide variety of different atoms and molecules, in processes where the atom/molecule bonds to the defects (see for instance References 3–6).

Whilst the interatomic C–C bonds in pristine graphene are completely covalent, the bonding between the graphene sheet and other atoms can be described as a mixture of ionic and covalent bonding. We will begin, however, by reviewing the hydrogenation of graphene in which C–H bonds form which are almost completely covalent.

11.2 Hydrogenation of graphene

11.2.1 Theoretical aspects

Fully hydrogenated graphene, named graphane ([1] of Chapter 1 and [7]), can be understood in simple terms as a graphene sheet in which each carbon atom is covalently bonded to a hydrogen atom in addition to the three adjacent carbon atoms. The bonding to an additional atom changes the orbital hybridization from sp^2 to sp^3 and causes the carbon atoms to move out of the graphene plane, allowing the necessary change in bond angles for sp^3 hybridization. The structure for graphane believed to be the most stable, the chair conformation, can be visualized as a one atom thick slice through the diamond lattice [8], with the covalently bonded hydrogen atoms terminating the broken bonds above and below the slice (Figure 11.1).

We expect that graphane can also exist in a variety of other conformations: the boat, twist-boat and twist boat–chair conformations. Figure 11.2 shows the two conformations expected to be most stable: the chair and boat conformations. In the chair conformation, the system is free to adopt 109.5° bond angles in a tetrahedral structure (ensuring the largest possible separation between the electron clouds forming adjacent bonds), and there is a large distance between the hydrogen atoms. We therefore expect that the chair conformation will not exhibit any steric strain; the system is free to adopt the bond angles that result in the strongest pure sp^3 bonds. This is not the case for the other conformations. It is for this reason that we expect the chair conformation to be the most stable conformation of graphane.

The results of more detailed theoretical calculations [9] are in agreement with our expectation on the basis of the arguments mentioned earlier. It is expected (Figure 11.3) that the graphene–molecular hydrogen system should save

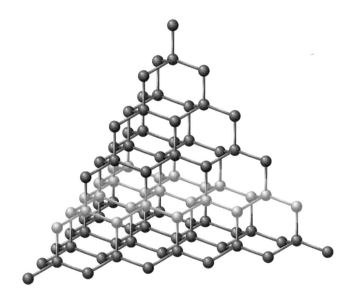

FIGURE 11.1
The diamond lattice. The chair conformation of graphane can be visualized as a one-atom thick slice through this lattice (highlighted) with hydrogen atoms terminating the broken bonds above and below the slice.

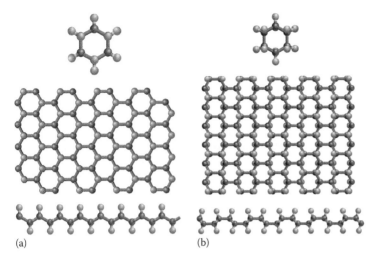

FIGURE 11.2
Chair (a) and boat (b) conformations of cyclohexane (top) and graphane (middle and bottom). Carbon atoms are shaded.

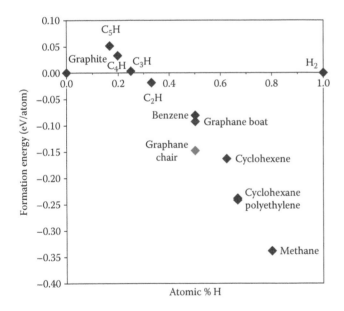

FIGURE 11.3
Formation energy per atom predicted using density functional theory as a function of hydrogen content for some graphane conformations and some molecular hydrocarbons for comparison. (Reprint with permission from Sofo, J.O., Chaudhari, A.S. and Barber, G.D., *Phys. Rev. B*, 75, 153401. Copyright 2007 by the American Physical Society.)

a considerable amount of energy when forming graphane in the chair formation, and that this material should be more stable than benzene (with the same stoichiometry).

11.2.2   Preparation of hydrogenated graphene and measurement of hydrogen content

In contrast to the simplicity of the graphene-hydrogen system from the theoretical point of view, graphane and hydrogenated graphene materials have proved extremely difficult to synthesize and accurately characterize. On a large (~1 µm+) scale, it is reasonable to state that it is not possible to accurately measure the hydrogen content in a hydrogenated graphene crystal; hydrogen content can only be estimated. Our ability to estimate hydrogen content is, however, good enough to conclude that the hydrogen content in hydrogenated graphene samples synthesized to date is nowhere near to 100 atomic percent (at.%); 10 at.% would be a more reasonable estimate.

The preparation of hydrogenated graphene* was first achieved by two different processes: bombardment of pristine graphene with atomic hydrogen in vacuum ([1] of Chapter 1) and electron-beam-initiated decomposition of a hydrogen silsesquioxane film covering the graphene sample [10]. The former method has since been more widely adopted. Hydrogenated graphene samples have been prepared from graphene on a conventional substrate (in which case only single-sided hydrogenation is likely to occur) and from graphene mounted on a substrate containing an aperture (in which case the hydrogen can access and bond to both sides of the sample).

A number of further studies followed, in which hydrogenated graphene samples were characterized using a variety of techniques. We will review here the key properties of hydrogenated graphene samples, and how we know (or are not certain of) these key properties.

Let us examine first the techniques used to estimate hydrogen coverage in hydrogenated graphene samples. We can begin with Raman spectroscopy; in pristine mono-layer graphene, the only first-order Raman peak permitted by the selection rules is the zone-centre optical phonon at 1583 cm^{-1} which gives rise to the G peak. But when we allow hydrogen to bond to a small proportion of the carbon atoms, pulling them out-of-plane, it is possible for the (first order) D peak to appear also.

---

* A number of publications use the term "graphane" to describe the synthesized hydrogenated graphene samples. Technically this is incorrect because "graphane" refers only to fully hydrogenated (100 at.%) graphene and the hydrogenated graphene samples synthesized to date do not even approach this C/H ratio.

The D peak is forbidden in pristine graphene because it does not originate from a zone-centre phonon so momentum cannot be conserved when it is optically excited. However, carbon atoms which have hydrogen attached are expected to act as defects. The electron excited to the conduction band in the double resonant Raman scattering process can scatter off one of these atoms just as it scatters off a defect. Thus the D peak becomes allowed by the selection rules following even a very small hydrogen content for the same reasons as it becomes allowed in defective graphene (Section 7.7).

Figure 11.4 shows Raman spectra of hydrogenated graphene samples prepared under conditions leading to different hydrogen content. It is possible to estimate the hydrogen content from the intensity ratio of the D and G peaks [11]. The methodology is as follows. We assume that, due to the much larger Raman scattering cross-section of sp^2-bonded carbon compared to sp^3-bonded carbon under visible or IR excitation ([1] of Chapter 1 and [14] of Chapter 7), the Raman signal observed originates from the parts of the sample which remain in their pristine sp^2-bonded form with no hydrogen attached. As we increase the hydrogen content, the diameter d of these parts of the sample decreases and the

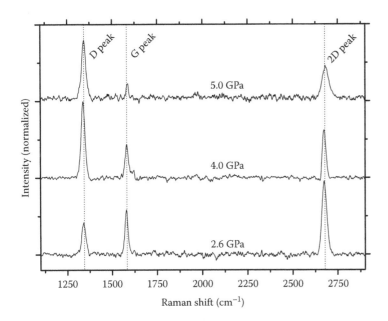

FIGURE 11.4

Raman spectra of hydrogenated graphene samples produced by reaction of graphene with hydrogen at 200°C under different hydrostatic pressures, leading to different hydrogen content. The highest D/G intensity ratio (5 GPa, top spectrum) corresponds to the highest hydrogen content. (Adapted from Smith, D. et al., *ACS Nano*, 9, 8279, 2015 [Chapter 10]. With permission.

D peak intensity I_D grows relative to the G peak intensity I_G just as it does for microcrystalline graphene when the crystallite size is decreased [12]:

$$d \approx 2.4 \times 10^{-10} \lambda^4 \times \frac{I_G}{I_D} \qquad (11.1)$$

It has been proposed [12] that on this basis an I_D/I_G ratio of two should correspond to a hydrogen coverage of 10 at.% but this should be regarded as a very rough estimate. The relation between hydrogen coverage and I_D/I_G ratio is probably not the same for hydrogenated graphene samples prepared using different methods. There are a number of reasons for this.

First, bonding of hydrogen to one side of the sample may induce the other side to bond to the substrate and the manner in which this occurs may vary according to the substrate used (SiO$_2$, Cu, SiC, etc.). Second, the relation between hydrogen coverage and the diameter d of the remaining areas of pristine graphene must vary depending on how evenly distributed over the sample the hydrogen atoms are. Third, the creation of (actual) defects in the graphene lattice also causes the D peak to appear. In some studies (for instance, References [1] of Chapter 1 and [49] of Chapter 10), the D peak disappears completely upon removal of hydrogen (see Section 11.2.6), indicating that its appearance can be attributed solely to the addition of hydrogen. In other studies (for instance Reference 12), some residual D peak intensity remains after removal of the hydrogen, suggesting that the creation of (actual) defects during the hydrogenation process also contributed to the observed D peak intensity.

In any case, we only expect the relation in Equation 11.1 to apply for small hydrogen content. For higher hydrogen content, we would expect the I_D/I_G ratio to decrease and, for both peaks to broaden significantly, the same as for larger defect concentration (Figure 7.11). Pristine graphane should have no measurable Raman spectrum, due to the extremely low Raman scattering cross-section of sp^3-bonded carbon compared to sp^2-bonded carbon under visible or IR excitation (it is roughly a factor of 50 lower) ([1] of Chapter 1 and [14] of Chapter 7). Whilst pristine graphane (i.e. with C/H ratio of 1:1) has not been prepared to date, pristine fluorographene (C/F ratio of 1:1) has been prepared ([34] of Chapter 9) (Section 11.3.2) and has no measurable Raman spectrum.

The other method commonly used to estimate hydrogen content in hydrogenated graphene samples is synchrotron X-ray photoelectron spectroscopy. This technique utilizes X-rays to measure the binding energy of the electrons in the 1s orbital of the carbon atoms. There is a small shift in binding energy depending on the environment of the carbon atom: part of an area of pristine graphene, adjacent to a carbon atom which is bonded to hydrogen or actually bonded to a hydrogen atom [13]. The width of these peaks is of similar size to the separation between the adjacent peaks so an accurate measurement of the relative intensities of the peaks and hence of hydrogen content is challenging. This is especially true given that the degree of hybridization change and intensity ratio between the various peaks should vary significantly depending on how evenly the hydrogen atoms are distributed over the graphene lattice.

11.2.3   Lattice constants of hydrogenated graphene

An ideal hydrogenated graphene sample is a sample suspended over a substrate with an aperture, not only because that allows better hydrogen coverage* but because it allows characterization of the sample using a greater range of experimental techniques. In particular, the access to the sample from both sides makes electron diffraction measurements of the hydrogenated graphene lattice constants possible.

Elias et al. ([1] of Chapter 1) performed a number of electron diffraction measurements on a sample of suspended hydrogenated graphene. The results (Figure 11.5) demonstrated a wide variation in lattice constant and hence a

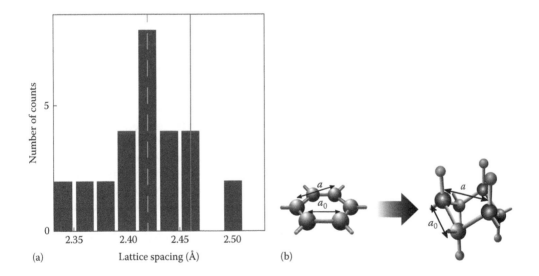

FIGURE 11.5

(a) Lattice constants of different areas of a hydrogenated graphene sample. Vertical line indicates value always observed for graphene ([1] of Chapter 1), with error of ± 0.01 Å ([34] of Chapter 9). (b) Schematic illustrating C–C bond length and lattice constant for graphene and graphane in the chair conformation. ([a]: From Elias, D.C. et al., *Science*, 323, 610, 2009 [Chapter 1]. Reprinted with permission of AAAS.)

---

* The impermeability of graphene to even the smallest atoms is well documented [14,15].

high level of disorder in hydrogenated graphene, implying spatially inconsistent hydrogen coverage. It is also noteworthy that a reduction of the lattice constant upon hydrogenation is observed in most measurements. This is a surprising result.

What change in the lattice constant do we expect to observe upon hydrogenation? There are two competing effects, as illustrated in Figure 11.5. Let us assume that graphane is formed in the most stable (chair) conformation. On the one hand, we expect the C–C bond length a_0 to increase to become close to the C–C bond length in the archetypal sp^3-bonded carbon system, diamond: 1.54 Å. This should lead to an increase in the lattice constant. On the other hand, we expect a change in structure from the graphene structure where all carbon atoms lie in the xy plane to the structure shown in Figure 11.5 where alternate carbon atoms lie above and below the xy plane with a tetrahedral structure and bond angles of 109.5°. This should lead to a decrease in the lattice constant.

Performing the necessary trigonometry using 1.54 Å, 109.5° leads us to the prediction that the lattice constant should increase from 2.46 Å to 2.51 Å upon hydrogenation. This primitive calculation is in agreement with the results of more detailed theoretical studies [9,16]. But in actual fact, as outlined earlier, the lattice constant decreases upon hydrogenation. The reasons for this are not well understood.

11.2.4   Disorder, frustration and steric strain in hydrogenated graphene

The electron diffraction measurements discussed in the previous section suggest that hydrogenated graphene samples exhibit a relatively high degree of disorder. This suggestion is confirmed by observations with scanning tunnelling microscopy (STM) [17,18] and scanning transmission electron microscopy [19]. These techniques cannot accurately estimate the overall hydrogen coverage due to the fact that an extremely small (on an atomic scale) area of the sample is imaged and also due to the difficulty of imaging individual hydrogen atoms. However, they can confirm that the regular, ordered lattice present in a pristine graphene sample is absent from hydrogenated graphene samples (Figure 11.6).

In principle, the pristine chair conformation of graphane as shown in Figures 11.1 and 11.2 should be a highly ordered structure with no steric strain so at first sight the observation of a high level of disorder in hydrogenated graphene is surprising. However, on reflection a highly disordered structure is what we expect. When we expose the graphene lattice to hydrogen on both sides, to obtain the pristine chair conformation of graphane, we need to somehow ensure that the A atoms in the graphene lattice only bond to hydrogen

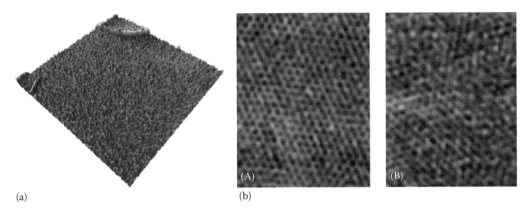

(a)   (b)

FIGURE 11.6
(a) Scanning tunnelling microscopy image (100 nm × 100 nm area) of hydrogenated graphene on a SiC surface. (b) High-angle annular dark field scanning transmission electron microscopy image of pristine (A) and hydrogenated (B) free-standing graphene membranes. In both cases the highly disordered nature of hydrogenated graphene samples is illustrated. ([a]: Reprinted with permission from Sessi, P., Guest, J.R., Bode, M. and Guisinger, N.P., *Nano Lett.*, 9, 4343. Copyright 2009 American Chemical Society; [b]: Reprinted with permission from Bangert, U., Pan, C.T., Nair, R.R. and Gass, M.H., *Appl. Phys. Lett.*, 97, 253118. Copyright 2010 American Institute of Physics.)

atoms on one side of the lattice and that the B atoms in the graphene lattice only bond to hydrogen atoms on the other side. Of course, there is no way to ensure this, so a frustrated lattice must result.

This is not the only obstacle to obtaining pristine ordered graphane. We discussed earlier the theoretical expectation that the graphene lattice should expand when hydrogenation takes place, and the experimental observation that it instead contracts. Whether the graphene lattice is seeking to expand or contract upon the addition of hydrogen, it is not free to do so because graphene is not stable in a completely free-standing form (Section 10.2) so the sample must be attached to a substrate. Even a graphene sample prepared over an aperture in the substrate must be firmly anchored down by van der Waals forces around the edges of the aperture.

For these reasons, we expect any real hydrogenated graphene sample to exhibit a high degree of disorder and steric strain.

11.2.5 Electronic properties of hydrogenated graphene

It is reasonable to state that the lack of an electronic bandgap in graphene is the principal obstacle to the realization of its potential applications in electronics. The lack of a bandgap ensures that a graphene transistor cannot be turned off in an effective manner. In simple terms, the effect of hydrogenation on the electronic properties of graphene is to open an electronic bandgap. However, it is important to understand this does not mean that hydrogenation removes the obstacle to the realization of graphene's electronics applications.

This is because hydrogenation opens a bandgap by removing the part of the dispersion relation responsible for the extremely high velocity of the electrons in graphene.

Were it possible to synthesize pristine graphane, we would expect it to be a wide bandgap insulator with electronic properties broadly similar to those of diamond. This is because graphane should exhibit pure sp^3 hybridization.

In sp^2 hybridization (graphene), it is the electrons which do not take part in hybridization (the π electrons) which are responsible for part of the dispersion relation where the conduction and valence bands meet at the K point and the extremely high conductivity. The other valence electrons are in the σ-bonds. The σ-bonds take a lot of energy to break and hence there is a large energy gap between the valence band (bonding orbitals) and conduction band (anti-bonding orbitals) for electrons in the σ-bonds.

In sp^3 hybridization (i.e. pristine graphane), all the valence electrons would be in strong σ-bonds, which is why we expect graphane to be a wide bandgap insulator similar to diamond. This expectation is borne out by more detailed theoretical calculations using density functional theory (Figure 11.7). We observe that the conduction and valence bands are closest at the Γ point, where an extremal bandgap of nearly 4 eV is

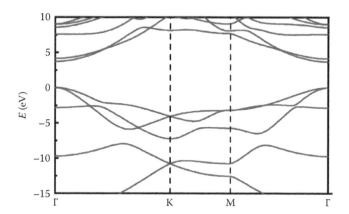

FIGURE 11.7

Calculated electronic dispersion relation of pristine graphane in the chair conformation. Fermi level is at 0 eV. (Reprinted with permission from Leenaerts, O., Peelaers, H., Hernández-Nieves, A.D., Partoens, B. and Peeters, F.M., *Phys. Rev. B*, 82, 195436. Copyright 2010 by the American Physical Society.

observed. The intrinsic conductivity resulting from a material with this dispersion relation would be extremely small ($(3/2)k_BT \sim 10^{-2}$ eV at ambient temperature). This is in contrast to pristine graphene, in which the conduction and valence bands for the π electrons meet at the K point.

For the hydrogenated graphene samples synthesized to date with relatively small hydrogen content (~10 at.%), we also expect a drastic reduction in conductivity compared to pristine graphene. This is because the extremely high conductivity of pristine graphene is a result of the ballistic transport exhibited by the material (Section 3.6); electrons can move extremely large distances without being scattered. Addition of even tiny amounts of hydrogen should result in movement of the carbon atoms bonded to hydrogen out of the graphene plane. These atoms will hence act as defects, scattering the electrons and preventing ballistic transport from taking place. Experimentally, a large decrease in conductivity for even small hydrogen content is observed ([1] of Chapter 1 and [11]) despite angle-resolved photoelectron spectroscopy (ARPES) evidence suggesting that the resulting electronic bandgap is relatively small [13].

11.2.6   Thermal stability of hydrogenated graphene

In general terms, the sp^3 C–H bond is an extremely strong bond. Methane (CH_4), for instance, is extremely resistant to thermal decomposition [21]. In contrast, many authors have observed that hydrogen is lost from hydrogenated graphene samples through vacuum annealing at much lower temperatures: 450°C ([1] of Chapter 1), 325°C [13], 500°C [12], 200°C ([49] of Chapter 10) and even 75°C [22].

This poor thermal stability is a barrier to many applications (for instance in electronics). But it enables other potential applications, as it ensures the hydrogen can be easily removed on demand or replaced by other atoms and molecules. It is difficult to see how, in a pristine freestanding layer of graphane in the chair conformation, the C–H bond would be any weaker than that in sp^3-bonded molecular hydrocarbons such as methane. Therefore, the most likely reason for the experimentally observed poor stability of hydrogenated graphene is that the poor hydrogen coverage, disorder and steric strain discussed earlier prevent the system from adopting pure sp^3 hybridization. The system is not free to adopt the bond lengths and angles necessary for the sp^3 molecular orbitals to form with the largest possible overlap between the hybridized atomic orbitals and the correct bond angles to ensure neighbouring electron clouds are kept apart; and hence the bonds are relatively weak.

11.3   Halogenation of graphene

11.3.1   General remarks: Ionic and polar covalent bonding

Pristine mono-layer graphene will also react readily with the halogen atoms towards the top of the periodic table: fluorine and chlorine. To understand the materials synthesized in these reactions, we must consider the partially ionic nature of the interatomic bonding in this case. Thus far, we have discussed C–C and C–H chemical bonds which are covalent and described effectively using molecular orbital theory. This means that they form because the atoms concerned can save some energy by sharing electrons (Chapter 2). The sp, sp^2 and sp^3 C–C bonds are completely covalent. If we drew a plane bisecting the bond axis, there would be no net accumulation of charge on either side of the plane. The sp^3 C–H bond is almost completely covalent. If we drew such a plane bisecting this bond, we may discover that the hydrogen atom has a very small net positive charge and the carbon atom has a very small net negative charge. So it is a reasonable approximation to view the bond as covalent.

In halogenated graphene materials, however, the bonding is not completely covalent. It is partially ionic in nature. As a result, the strength of the bond is partially due to the energy saving when the electrons are shared, and partly due to the electrostatic (Coulomb) attraction between the positively charged carbon atom and the negatively charged halogen atom.

To understand the mixed ionic and covalent nature of the bonding in halogenated graphenes, we need to understand the factors which make formation of covalent and ionic bonds energetically favourable.

In the case of ionic bonds, we use a parameter called the electronegativity of the relevant atoms. The electronegativity of an atom is a dimensionless parameter encapsulating various pieces of information about the energy cost or saving when it gains and loses an electron. Various electronegativity scales are in use. Here, we will use the Pauling scale ([4] of Chapter 2) [23] (Table 11.1). The larger the difference in electronegativity between the two bonding atoms, the greater the energy saving when an ionic bond is formed.

On the other hand, the energetic favourability of covalent bond formation depends on the degree of overlap which is possible between the atomic electron wave functions when forming the molecular orbital. Thus, for instance, hydrogen forms a much stronger covalent bond to carbon than to silicon because the hydrogen electronic wave function (the spatially small 1s wave function) can overlap with a much larger proportion of the carbon sp^3 hybridized wave function than with the silicon sp^3 hybridized wave function. Hence, whilst methane (CH_4) is stable in air at ambient conditions, does

TABLE 11.1
Pauling electronegativities of selected light elements

H						
2.20						
Li	Be	B	C	N	O	F
0.98	1.57	2.04	2.55	3.04	3.44	3.98
Na	Mg	Al	Si	P	P	Cl
0.93	1.31	1.61	1.90	2.19	2.58	3.16
K	Ca	Ga	Ge	As	Se	Br
0.82	1.00	2.01	2.01	2.18	2.55	2.96

Source: Atkins, P.W., *Physical Chemistry*, OUP, 1982; Allred, A.L.,
       J. Inorg. Nucl. Chem., 17, 215, 1961 [Chapter 2].

not thermally decompose easily [21] and is stable at ambient temperature under high pressure until at least 200 GPa pressure [24], silane (SiH_4) is pyrophoric and decomposes readily under a variety of conditions ([21] of Chapter 8).

11.3.2   Fluorination of graphene

Out of all the halogens fluorine has the highest electronegativity, enabling it to form a stronger ionic bond to carbon than other halogens. In addition, its valence electrons originate from the same principal shell ($n = 2$) as those in carbon enabling the formation of a molecular orbital with the largest possible overlap between the atomic electron wave functions and hence the strongest possible covalent bond to carbon.

So whatever the character of the bonding in fluorinated graphene, we expect graphene to bond more strongly to fluorine than to any of the other halogens. The experimental evidence is consistent with this. Whilst graphene is stable when exposed to molecular F_2, in the presence of atomic fluorine (provided by the decomposition of XeF_2 at $\approx 70°C$) it reacts readily ([34] of Chapter 9). Within the experimental error, the fluorine coverage is 100 at.%. This is reflected in the disappearance of the Raman spectrum (as we would expect in the case of a complete conversion to sp^3 bonding). Figure 11.8 demonstrates the progressive disappearance of the Raman spectrum of graphene upon increasing the fluorination level to 100 at.% along with an electron diffraction pattern from the fully fluorinated sample. The diffraction pattern indicates that the hexagonal symmetry of the graphene lattice is preserved; that is, the sample is fluorographene rather than an amorphous mixture of carbon and fluorine.

We do however expect fully fluorinated graphene to exhibit a high degree of structural disorder. Earlier, we discussed the reasons why it should not be possible experimentally to form a pristine crystal of graphane in the chair

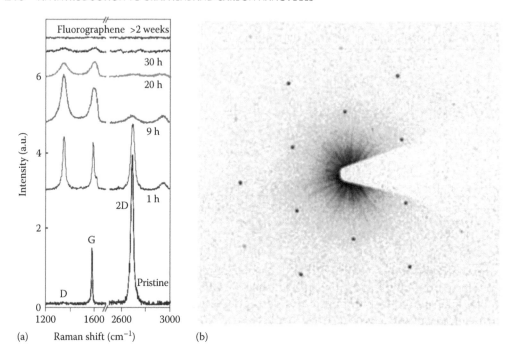

FIGURE 11.8
(a) Raman spectra of graphene following exposure to atomic fluorine for different time periods, providing increasing levels of fluorination up to fully fluorinated graphene (top spectrum). (b) Electron diffraction pattern of a fully fluorinated graphene sample demonstrating the preservation of the graphene lattice. (From Nair, R.R.: *Small*. 2010. 6. 2877 [Chapter 9]. Copyright Wiley-VCH Verlag GmbH & Co. KGaA. Reproduced with permission.)

conformation: the fact that it would require the hydrogen atoms to "know" that they were only supposed to bond to the A carbon atoms on one side of the graphene plane and only to the B atoms on the other side. This argument applies equally to the reaction between fluorine and graphene; we do not expect a large single crystal of fluorographene in the chair (or any other) conformation to form. Instead there must be a high degree of structural disorder and lattice frustration.

In contrast to hydrogen, fluorine is very hard to remove from graphene once it has been attached. Hydrogen is lost from graphene under annealing to just 75°C in vacuum [22], whilst the removal of fluorine from fully fluorinated graphene requires prolonged annealing in an argon–hydrogen mixture at 400°C, although partially fluorinated graphene is less stable ([34] of Chapter 9).

11.3.3 Other halogenation reactions

Whilst fluorographene has been synthesized with approximately 1:1 stoichiometry, other halogenation reactions have been less successful. Chlorination has been achieved using a photochemical process [25] and a plasma process [26] but coverage does not approach 100 at.%. Bromographene is not expected to be stable [27] and bromographene has not been prepared from a pristine mono-layer graphene sample. This is to be expected, since there are a number of factors that should ensure that graphene cannot react with larger halogens to form a stable halogenated graphene. The formation of ionic carbon–halogen bonds becomes less energetically favourable for larger halogens due to the lower difference in electronegativity. The formation of covalent carbon–halogen bonds becomes less energetically favourable due to the smaller degree of overlap between the atomic electron wave functions, and since the halogen atoms lower down the periodic table become very large compared to the carbon atom there may not be adequate room for them above and below the graphene lattice to achieve a high coverage.

11.3.4   Electronic dispersion relation of halogenated graphenes

We expect the bonding in halogenated graphene samples to be a mixture of covalent and ionic bonding. Earlier, we discussed the reasons why the formation of a covalent bond and consequent change to sp^3 hybridization should result in the drastic loss of electrical conductivity. The same is true for ionic bonding of atoms to a graphene layer; in a pristine graphene layer it is the atoms in the unhybridized $2p_z$ orbitals which are the most weakly bound. Hence if an atom bonds ionically to a graphene layer, it is these electrons (the ones responsible for the high conductivity) which would be lost. Hence, we expect halogenated graphenes to be insulators regardless of whether the bonding is covalent, ionic or polar covalent. Experimentally, both fluorographene ([34] of Chapter 9) and chlorographene [25] have turned out to be insulators.

11.4   Graphene oxide

Thus far, we have discussed the reaction of graphene with hydrogen and halogens, all with a valency of 1, but the chemistry of graphene oxide is fundamentally different. There is a stable periodic structure without steric strain (the chair conformation) which exists for both hydrogenated and halogenated graphene, even though it may be challenging to attain experimentally. Oxygen, however, has a valency of 2 so there is no stable strain-free periodic structure which exists for graphene oxide.

   In addition, we can prepare hydrogenated and halogenated graphene samples with no impurities present in significant quantity. This is not possible for graphene oxide. The materials referred to as "graphene oxide" in the literature are in fact highly disordered materials (sometimes even described as amorphous [28]) in which oxygen, hydrogen, hydroxyl groups and other impurities are all bonded to the graphene lattice [28–30]. The bonding is therefore a poorly characterized mixture of sp^2 and sp^3 hybridization.

For these reasons, we will not discuss graphene oxide materials further in this chapter. But it is worth noting that, despite the high level of structural disorder, samples of bulk few-layer graphene oxide (obtained by exfoliation of graphite oxide) are one of the most promising graphene-related materials in terms of potential applications. These are discussed in Chapter 12 and also in a number of review papers in the literature (for instance, References 28–30).

References

1. L Liu et al., *Nano Lett.* **8**, 1965 (2008).

2. F Karlický, KKR Datta, M Otyepka and R Zbořil, *ACS Nano* **7**, 6434 (2013).

3. D Chen, H Feng and J Li, *Chem. Rev.* **112**, 6027 (2012).

4. HL Poh, S Sofer, M Nováček and M Pumera, *Chemistry* **20**, 4284 (2014).

5. HL Poh, P Šimek, Z Sofer and M Pumera, *ACS Nano* **7**, 5262 (2013).

6. L Wang, Z Sofer, P Šimek, I Tomandl and M Pumera, *J. Phys. Chem. C* **117**, 23251 (2013).

7. KS Novoselov, *Beyond the Wonder Material*, Physics World (August 2009).

8. A Savchenko, *Science* **323**, 589 (2009).

9. JO Sofo, AS Chaudhari and GD Barber, *Phys. Rev. B* **75**, 153401 (2007).

10. S Ryu et al., *Nano Lett.* **8**, 4597 (2008).

11. BR Matis, JS Burgess, FA Bulat, AL Friedman, BH Houston and JW Baldwin, *ACS Nano* **6**, 17 (2012).

12. JR Burgess et al., *Carbon* **49**, 4420 (2011).

13. D Haberer et al., *Nano Lett.* **10**, 3360 (2010).

14. JS Bunch et al., *Nano Lett.* **8**, 2458 (2008).

15. S Hu et al., *Nature* **516**, 227 (2014).

16. MZS Flores, PAS Autreto, SB Legoas and DS Galvao, *Nanotechnology* **20**, 465704 (2009).

17. P Sessi, JR Guest, M Bode and NP Guisinger, *Nano Lett.* **9**, 4343 (2009).

18. R Balog et al., *JACS* **131**, 8744 (2009).

19. U Bangert, CT Pan, RR Nair and MH Gass, *Appl. Phys. Lett.* **97**, 253118 (2010).

20. O Leenaerts, H Peelaers, AD Hernández-Nieves, B Partoens and FM Peeters, *Phys. Rev. B* **82**, 195436 (2010).

21. P Lenz-Solomun, M-C Wu and DW Goodman, *Catal. Lett.* **25**, 75 (1994).

22. Z Luo et al., *ACS Nano* **3**, 1781 (2009).

23. AL Allred, *J. Inorg. Nucl. Chem.* **17**, 215 (1961).

24. L Sun et al., *Chem. Phys. Lett.* 473, **72** (2009).

25. B Li et al., *ACS Nano* **5**, 5957 (2011).

26. X Zhang et al., *ACS Nano* **7**, 7262 (2013).

27. L Liao, H Peng and Z Liu, *JACS* **136**, 12194 (2014).

28. DR Dreyer et al., *Chem. Soc. Rev.* **39**, 228 (2010).

29. DR Dreyer, AD Todd and CW Bielawski, *Chem. Soc. Rev.* **43**, 5288 (2014).

30. G Eda and M Chhowalla, *Adv. Mater.* 22, 2392 (2010).

12 Current Topics in Graphene and Carbon Nanotube Research

12.1 Overview of the potential applications of carbon nanotubes and graphene

Measurements of the properties of individual single-walled carbon nanotubes (SWCNTs) and graphene flakes have yielded some superlative findings, such as being the strongest materials, the best conductors of electricity and heat, flexible and bendable, transparent, etc. Translating these properties to real-world applications, however, requires overcoming several challenges. In the case of electronic device applications, such challenges centre on control of the device's electronic properties as well as wafer-scale fabrication and reliability. For composites, the challenges lie in harnessing the nanoscale 1D and 2D mechanical, electrical and thermal properties and constructing enhanced useful macro-scale 3D structures. In the case of biomedical applications, the toxicology and bio-distribution of these materials is still hotly debated. In this chapter, we will look at some of the applications where CNTs and graphene show the greatest potential and discuss the obstacles that scientists are striving to overcome to make these applications a reality.

12.2 Electronics applications of carbon nanotubes and graphene

12.2.1 Carbon nanotube based electronic devices

As we have learnt before, SWCNTs exist in a variety of flavours, or chiralities, each with their unique electronic properties. This is both an advantage and a disadvantage in the application of SWCNTs in electronics. On one hand, it allows for various different electronic components such as transistors and interconnects to be fabricated from different SWCNT flavours, making an "all-carbon" electronic circuit possible. On the other hand, chirality or even electronic-type (metallic or semiconducting) specific synthesis of SWCNTs has proven elusive. When SWCNTs are grown with mixed chiralities, it poses a major challenge in placing specific kinds of SWCNTs at specific locations in a circuit on wafer scale to realize the aforementioned all-carbon electronics.

In Chapter 9, we have learnt about some ongoing efforts in chirality-selective growth of SWCNTs and the alternative approach to sort SWCNTs by their chirality or electronic type after growth. The latter approach has moved forward in leaps and bounds over the past few years. The sorting is typically performed on a dispersion of SWCNTs, by fractioning the dispersion using techniques such as density-gradient ultracentrifugation, gel electrophoresis, etc. [1].

In one approach, covalent and non-covalent functionalization chemistries can be used to discriminately affect SWCNTs in solution, thereby sorting them according to electronic type, diameter, chiral angle, and chiral handedness. For instance, fluorine-based chemistries can yield metal-semiconductor separation and diameter selectivity, flavin mononucleotides and polyaromatic amphiphiles are strongly selective of specific chiral angles, and customized mono- and di-porphyrins can discriminate between left- and right-handed chiralities. Another effective and widely used chemical sorting technique is selective DNA wrapping. Recently, over 20 DNA sequences have been identified which select for a specific chirality and can result in near perfect sorting for that specific chirality (Figure 12.1a) [2].

Another popular approach is density gradient ultracentrifugation (DGU), which is a bioinspired technique whereby the SWCNTs are sorted according to their buoyant density [3]. SWCNTs are set in motion under centrifugal force within a density gradient that is intentionally formed in a centrifuge tube. The SWCNTs come to rest at their respective isopycnic points (i.e. the location where the buoyant density of the SWCNT matches the local density of the gradient). The buoyant density of the SWCNTs depends on, and can be modified by, the type of density gradient medium, choice of solvent, choice of surfactant or surfactant mixture, etc. Variations of the DGU technique can also be used to sort SWCNTs by chiral angle and length, and also to sort double-walled CNTs according to metallicity of the walls.

A third sorting technique that has gained recent prominence uses agarose gel electrophoresis, which whilst resulting in lower purity of sorting between metallic and semiconducting SWCNTs, has the advantage of being an experimentally simple technique [4]. The ability to sort metallic SWCNTs by diameter has led to semi-transparent conductive films with tunable optical absorbance in the visible and infrared regions of the electromagnetic spectrum. Enrichment with semiconducting SWCNTs has led to thin-film field-effect transistors (FETs) with high switching ratios ($\sim 10^3$) and high on-state currents (>1 mA), delivering strong photocurrents, photoluminescence and electroluminescence.

A second challenge in realizing SWCNT electronics is the selective placement of specific types of SWCNTs onto the desired location and with desired orientation on a wafer scale. One approach which has proven particularly successful in this regard is bottom-up assembly by dielectrophoresis (DEP) [5]. DEP is the motion of polarizable particles relative to the fluid medium that they are dispersed in, under the influence of an inhomogeneous electric field. Depending on the

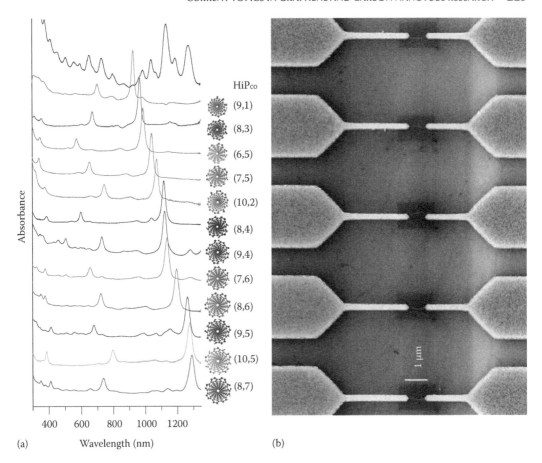

(a)   Wavelength (nm)   (b)

FIGURE 12.1

(a) Optical absorbance spectra of 12 single-chirality single-walled carbon nanotube (SWCNT) samples produced by ion-exchange chromatography of DNA-wrapped SWCNTs, compared to an as-synthesized HiPCO sample. (b) Individual SWCNTs assembled into device locations by bottom-up dielectrophoresis. ([a]: Reprinted by permission from Macmillan Publishers Ltd. *Nature*, Tu, X., Manohar, S., Jagota, A. and Zheng, M., 460, 250, copyright 2009; [b]: Reprinted with permission from Nano Letters, Vijayaraghavan A, Blatt S, Weissenberger D, Oron-Carl M, Hennrich F, Gerthsen D, Hahn H, Krupke R. 7, 6. Copyright 2007 American Chemical Society.)

relative polarizability between the particle and surrounding medium, the particle can undergo positive or negative DEP, wherein the particle moves towards or away from the region of higher field density. Preferential deposition of films or individual metallic SWCNTs has been demonstrating by taking advantage of intrinsic polarizability difference between metallic and semiconducting SWCNTs. However, deposition of semiconducting SWCNTs whilst rejecting metallic SWCNTs from a mixed dispersion directly using DEP has not yet been possible. In order to overcome this limitation, dispersions which are pre-sorted for semiconducting SWCNTs and even single-chirality SWCNTs has been used as the starting material, which has been assembled into arrays of devices of arbitrary positions and orientations.

Interestingly, there exists a mechanism by which when one SWCNT deposits in a device location, it modifies the local electric field distribution and changes the DEP force field around it from being attractive to repulsive, and keeping away additional SWCNTs from deposition. This self-limits DEP deposition to a single SWCNT per device location (Figure 12.1b). Using special electrode configurations such as floating capacitively coupled electrode arrays, ultra-large scale integration densities exceeding one million field effect transistors per cm^2 have been demonstrated. Based on these results, Steiner et al. at IBM [6] have evaluated the radio-frequency performance of SWCNT array transistors comprised of highly separated, semiconducting CNTs assembled by DEP on a fully scalable device platform. The extrinsic current gain in these devices was comparable to the state-of-the-art, and the extrinsic power gain was improved. The de-embedded, intrinsic current gain and power gain cut-off frequencies of 153 and 30 GHz are the highest values experimentally achieved to date.

12.2.2   Graphene-based electronic devices

In order for graphene to be applied in electronic devices, the electronic conductivity of graphene should be switchable, that is, have two conductivity states representing ON (conducting) and OFF (non-conducting). Whilst the conductivity of graphene can be modulated using an external electric field, a fully OFF state does not exist intrinsically in graphene due to the conduction and valence bands meeting at the K point (as discussed in Chapter 3). Therefore, various routes have been explored to engineer a fully OFF state in graphene.

One route to opening an electronic bandgap in graphene is to exploit lateral confinement of charge carriers in a graphene nanoribbon, which creates an energy gap near the K point. Graphene nanoribbons of varying widths and different crystallographic orientations have been fabricated by lithographic patterning of mono-layer exfoliated graphene (Figure 12.2a) [7]. An energy gap is observed for narrow ribbons, which scales inversely with ribbon width. Energy gaps in excess of 100 meV were observed for widths less than 20 nm [8], which could have potential technological relevance. Graphene nanoribbons could be compared to SWCNTs in terms of the confinement-induced bandgap, but there are some significant differences. Graphene nanoribbons are easier to fabricate on a wafer-scale using conventional top-down lithography whilst unconventional bottom-up methods have to be adopted for wafer-scale carbon nanotube assembly. However, significant bandgaps around 1 eV (relevant to opto-electronics applications) can only be obtained in SWCNTs whilst the bandgaps in graphene nano-ribbons are practically limited to about 200 meV.

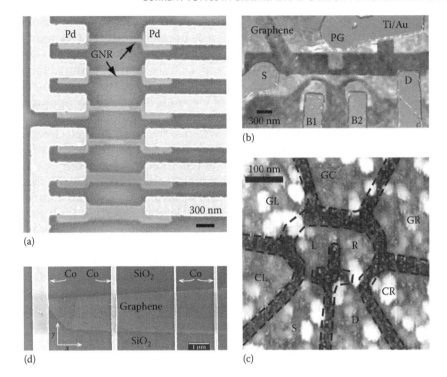

FIGURE 12.2
Scanning electron microscope images of (a) lithographically patterned graphene nanoribbons of different widths. (b) Graphene single quantum dot (QD). (c) Graphene double QD. (d) Graphene spin valve in non-local geometry. ([a]: Reprinted from *Physica E*, 40, Chen, Z., Lin, Y.-M., Rooks, M.J. and Avouris, P., 228, Copyright 2007, with permission from Elsevier; [b]: Reprinted with permission from Stampfer, C., Schurtenberger, E., Molitor, F., Güttinger, J., Ihn, T. and Ensslin, K., *Nano Lett.*, 8, 2378. Copyright 2008 American Chemical Society; [c]: Reprinted with permission from Molitor, F., Droscher, S., Guttinger, J., Jacobsen, A., Stampfer, C., Ihn, T. and Ensslin, K., *Appl. Phys. Lett.*, 94, 222107. Copyright 2009, American Institute of Physics; [d]: Reprinted by permission from Macmillan Publishers Ltd. *Nature*, Tombros, N., Jozsa, C., Popinciuc, M., Jonkman, H.T. and Wees, B.J.V., 448, 571, Copyright 2007.)

It is also possible to fabricate quantum dot (QD) devices entirely out of graphene using a lithographic procedure [9]. Such a device consists of a graphene island connected to the source and drain via two narrow graphene constrictions and three fully tunable graphene lateral gates as can be seen in Figure 12.2b. Larger QDs (>100 nm) show conventional single-electron transistor characteristics, with periodic Coulomb blockade peaks [10]. This can be extended to a double QD system [11] where the coupling of the dots to the leads and between the dots is tuned by graphene in-plane gates as can also be seen in Figure 12.2c. This structure has been proposed for the realization of spin qubits from graphene QDs.

Spintronics is the science and technology of manipulating electron spins rather than electron counts – in electronics the

ON and OFF conditions are related to having more and less electron flow, whereas in spintronics, the ON and OFF could, for instance, correspond to an electron current carrying predominantly UP and DOWN spin electrons. Flipping spins require significantly lower energy than reducing or increasing the number of electrons and therefore spintronics could be the future of low-power computing. The roadblock is that in most metallic systems, the spin of an electron is not preserved for very long distances, or in other words, the spin coherence length and lifetime are very short.

When graphene devices are fabricated with ferromagnetic electrodes, such as the soft magnetic Ni-Fe alloy or Co, it is possible to inject spin-polarised current into the graphene [12]. As discussed in Chapter 3, high-quality graphene enjoys ballistic transport with spin relaxation lengths between 1.5 and 2 μm even at room temperature, which is only weakly dependent on charge density. The switching fields of the electrodes can be controlled by in-plane shape anisotropy. Graphene spin valves have been constructed using either local or non-local geometry, and clear bipolar spin signal has been observed which is indicative of the relative magnetization directions of the electrodes, and magnetoresistance of up to 12% has been reported in local geometry. In Figure 12.2d, we see the scanning electron micrograph of a four-terminal single-layer graphene spin valve.

12.2.3   Graphene-based heterostructure electronic devices

Graphene devices on standard SiO_2 substrates are highly disordered, exhibiting characteristics that are far inferior to the expected intrinsic properties of graphene. Hexagonal boron nitride (hBN), on the other hand, offers an atomically smooth surface that is relatively free of dangling bonds and charge traps, a lattice constant similar to that of graphite and a large electrical bandgap, which allows graphene to perform close to its pristine nature. hBN/graphene layered heterostructures are produced by sequential exfoliation and transfer technique [13]. In the gate voltage dependence of conductivity of a graphene device on h-BN, the resistivity peak corresponding to the overall charge neutrality point is extremely narrow and occurs at nearly zero gate voltage. A carrier mobility of 60,000 cm^2 V^{-1} s^{-1} has been obtained, which is three times larger than that on SiO_2. Additionally, annealing such devices in a H_2/Ar flow at 340 °C for 3.5 h substantially enhances the carrier mobility whilst leaving the position of the charge neutrality point virtually unchanged.

An alternate method to overcome the lack of an OFF state in graphene is to construct a bipolar field-effect transistor that exploits the low density of states in graphene and its one-atomic-layer thickness. These devices comprise graphene heterostructures with atomically thin hBN or molybdenum disulphide (MoS_2) acting as a vertical transport barrier [14]. In Figure 12.3, we see a schematic representation of such a tunnelling field effect transistor [15]. Two graphene layers are separated by a hBN layer, and additional hBN layers exist on either side of the graphene layers to provide encapsulation and high mobility.

When a gate voltage V_G is applied between the Si substrate and the bottom graphene layer (Gr_B), the carrier concentrations n_B and n_T in both bottom and top electrodes increases because of the weak screening by mono-layer graphene. The increase of the Fermi energy E_F in each graphene layer leads to a reduction in barrier height for electrons tunnelling at this energy. Furthermore, the increase in the tunnelling density of states as the Fermi energy moves away from the neutrality point leads to an increase in the tunnel current. A bias voltage V_b applied between Gr_B and Gr_T gives rise to a tunnel current through the thin hBN barrier that scales with device area, leading to the ON state of the device.

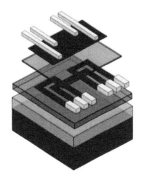

FIGURE 12.3

Schematic of a graphene/h-BN heterostructure device. Two graphene mono-layers (dark grey) are interlaid with hBN crystals (grey) and have separate electrical contacts (light grey). The heterostructure is fabricated on top of an oxidised silicon wafer. (Reprinted by permission from Macmillan Publishers Ltd. *Nat. Mater.*, Haigh, S.J. et al., 11, 764, Copyright 2012.)

12.3   Carbon nanotube and graphene composites

12.3.1   Composites with graphene materials

Using graphene oxide (GO) as a starting point, a number of avenues exist for the composition of graphene-based composites. For instance, the GO can be either used as such or reduced to form a reduced (r)GO or chemically modified to form a functionalized GO. The graphene material and the polymer material can be blended in solution phase in a mutual solvent, by adding the graphene to a melt of the polymer or by mixing the graphene material with the monomers followed by *in situ* polymerization. These are the three most commonly used methods for incorporation of the nanoscale reinforcements into polymers.

As an example, we will consider some combinations of these routes in achieving a graphene/polyurethane (PU) composite material [16]. In this example (Figure 12.4), GO produced by a modified Hummer's method was used as the starting point. The GO was either thermally reduced by rapid heating to 1050 °C to expel most of the oxygen functional groups or reacted with an isocyanate to form a suspension of functionalized GO which is stable in polar aprotic organic solvents. The matrix polymer used here was thermoplastic PU. Next, GO, thermally reduced GO, and isocyanate-functionalized GO were incorporated into a thermoplastic PU elastomer via melt intercalation, solvent mixing, and *in situ* polymerisation respectively.

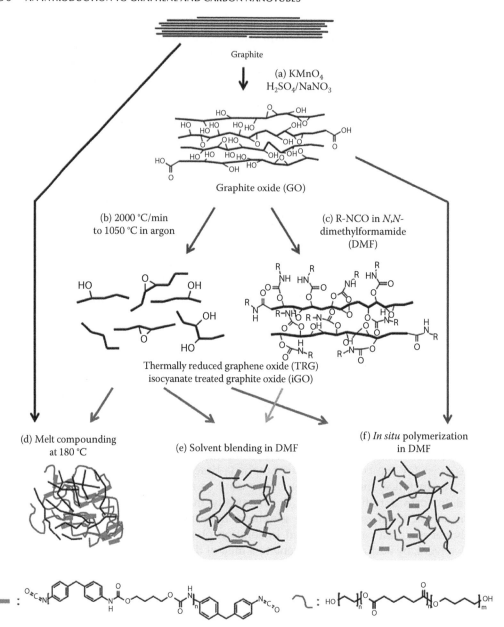

FIGURE 12.4
Schematic diagram of various routes for the production of a graphene-reinforced poly-urethane composite. (a) Oxidation of graphite. (b) Production of reduced graphene oxide. (c) Production of functionalised graphene oxide dispersion. (d) Melt compounding. (e) Solvent blending. (f) *In situ* polymerization. (Reprinted with permission from Kim, H., Miura, Y. and MacOsko, C.W., *Chem. Mater.*, 22, 3441. Copyright 2010 American Chemical Society.)

It is also, of course, possible to incorporate pristine or unfunctionalized graphene into a polymer matrix, although (as we saw in Chapter 9) it is challenging to produce large quantities of mono-layer graphene using solution-phase exfoliation, so invariably the unfunctionalized graphene

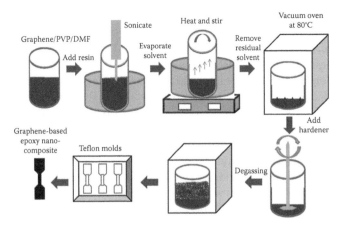

FIGURE 12.5
Schematic diagram of the steps involved in preparing a graphene/epoxy composite. (From Wajid, A.S., Tanvir Ahmed, H.S., Das, S., Irin, F., Jankowski, A.F. and Green, M.J.: *Macromol. Mater. Eng.* 2013. 298. 339. Copyright Wiley-VCH Verlag GmbH & Co. KGaA. Reproduced with permission.)

composites contain few-layer graphene or nanographite particles rather than mono-layer graphene as the reinforcement.

In another example [17] (Figure 12.5), pristine graphene was used as the filler in an epoxy matrix. The graphene itself is unstable in most organic and aqueous solvents as mentioned before, so in this case, polyvinylpyrrolidone (PVP) was used as a stabilizing agent to aid in the dispersion of the graphene. This effectively results in a graphene/PVP/epoxy composite. Another distinguishing feature here is that the preparation of the epoxy requires a hardener and chemical curing. The solvent used in this case was dimethylformamide (DMF) which disperses both the epoxy resin and the PVP-stabilized graphene.

The resulting composite had a 38% increase in strength and a 37% improvement in modulus for 0.46 vol.% graphene loading. In addition to the mechanical reinforcement, graphene additives are often used to provide electrical and thermal conductivity to the composite. The composites in this case had a very low electrical percolation threshold of 0.088 vol.%. Percolation threshold is the fraction of graphene or additive above which they form a long-range connected network through the polymer matrix, allowing for electrical and also thermal conductivity across a macroscopic sample.

Another kind of polymer system is a latex. A latex is a stable dispersion or emulsion of polymer microparticles in an aqueous medium. Whilst latex generally refers to the naturally occurring milky fluid in many flowering plants, synthetic latexes have also been made, for example of polypropylene (PP). PP latex is an aqueous emulsion which is composed of 30 wt.% of PP and 5 wt.% of oleic acid-based anionic surfactant, neutralised by an amino alcohol. Composites were fabricated by mixing PP latex with GO and a subsequent *in situ* chemical reduction of GO with hydrazine to yield a rGO and PP latex composite [18]. Whilst typically rGO would not be stable in water, the presence of latex particles and the associated anionic surfactant helps to steric and charge stabilize the *in situ* reduced GO. Finally, the rGO/PP composite was obtained by filtration of the black mixture through a 0.45 mm nylon membrane and washed with an excessive amount of water to remove the surfactants and hydrazine.

Natural biomaterials have drawn attention as a choice for nanocomposite matrices because of their inspiring morphologies, biocompatibility, biodegradability, and superior mechanical performance such as those demonstrated by silk materials. Nanocomposite membranes have been produced by incorporating GO sheets into silk fibroin matrix through heterogeneous surface interactions in an organised layer-by-layer assembly [19]. The multi-block nature of silk backbones with alternating hydrophilic and hydrophobic nanoscale domains facilitates a combination of hydrogen bonding, polar–polar and hydrophobic–hydrophobic interactions at the interfaces with GO, which also contains a similar distribution of domains. The Young's modulus of the nanocomposite membranes improved from about 10 GPa for pure silk fibroin to about 150 GPa for 20 vol.% GO. Similarly, the ultimate stress improved from about 100 to 300 MPa and the toughness from 0.7 to 2.2 MJ m^{-3}.

12.3.2   Composites with carbon nanotubes

Carbon nanotubes (CNTs – both single-walled and multi-walled) are an excellent filler material for nanocomposites, adding strength, electrical conductivity and thermal conductivity to any matrix [20]. The most common types of CNT composites, based on the type of matrix material, are CNT/polymer, CNT/ceramic and CNT/metal composites.

In order to maximize the impact of CNTs in a matrix, the CNTs have to be well dispersed, possibly aligned and the fibre/matrix interaction has to be tailored according to the desired properties. CNTs can be incorporated into a polymer matrix by processes such as solution mixing, *in situ* polymerization, melt compounding and chemical grafting processes. CNTs have been incorporated into a wide variety of polymer matrices such as polyethyleneterephthalate, polystyrene, PP, poly(methyl methacrylate), polycarbonate, poly(vinyl alcohol), polyaniline (PANI), polyethyleneimine, Nylon 6, polyimide, PU and epoxy resin. CNT composites with tensile strength and Young's modulus comparable to carbon fibre composites have been produced, ideally suited for structural applications. The electrical conductivity of the composites can be enhanced at a very low percolation threshold (<0.1 wt.%) of CNTs, which could be used for electrostatic dissipation, electromagnetic shielding applications and conductive coatings. Furthermore, it has been reported that generally, thermal conductivity and thermal stability of a polymer matrix is improved upon addition of CNTs.

Ceramic materials possess high stiffness, excellent thermal stability and low density, but are often brittle with low breaking strength. Addition of CNTs to ceramic matrices improves their toughness, creep resistance and thermal stability, as well as electrical and thermal conductivity. CNTs can be dispersed into a ceramic matrix, such as silica or alumina, but *in situ* growth, the sol-gel technique, spark plasma sintering, hot pressing, etc. of CNTs has also been used to reinforce metal matrices, such as aluminium, nickel, copper and magnesium. Powder metallurgy techniques such as mechanical alloying and sintering, mixing and hot pressing, spark plasma sintering as well as electrodeposition, electroless deposition, melting and solidification have been used for preparing CNT metal–matrix composites. Increase in tensile strength, hardness and wear resistance has been reported for CNT-reinforced metals. However, electrical and thermal conductivity of metals typically reduced upon the inclusion of CNTs, and the thermal expansion coefficient typically decreases. The production of both ceramic and metal–matrix composites poses similar challenges associated with high-temperature processing, such as homogeneous dispersion of CNTs in the matrices, interfacial bond strength between the two phases, and the chemical and structural stability of CNTs.

12.4 Biomedical applications of carbon nanotubes and graphene

12.4.1 Biocompatibility and biomedical applications of carbon nanotubes

Depending on the mode of interaction with the CNT, whether the CNTs are inhaled or taken into the vascular system, the resultant toxic effects of CNTs vary [21,22]. When CNTs are inhaled as dry particles or enter the respiratory system dispersed in aqueous media, studies have consistently reported inflammatory response, granuloma formation, progressive interstitial fibrosis and alveolar wall thickening in the lungs. This is predominantly due to the propensity of CNTs in dispersion to agglomerate; even when they are stabilized using surfactants and shortened and well-dispersed by sonication to begin with. Therefore, the actual toxicological effect of an individual CNT is overshadowed by the effects of CNT agglomerates which invariably form in the lungs. The accumulation and clearance kinetics further depends on the length and diameter of the CNTs, which is also linked to ease of agglomeration. This suggests that significant care must be taken in CNT production and handling to avoid the generation of airborne CNT particles which might be inhaled.

Cytotoxicity studies suggest that exposure to CNTs results in increased oxidative stress (increased free radical and peroxide generation and depletion of total antioxidant reserves), loss in cell viability and morphological effects on cellular structure. In early studies, this can be attributed to the high level of carbonaceous residues (soot) and catalyst nanoparticles of iron, nickel, etc., which are residual from the production processes due to inadequate purification. Later, it was shown that purified MWCNTs had a dose-dependent cytotoxicity similar to asbestos or carbon black. In this work, it was also noticed that unlike the pulmonary response, the agglomeration of CNTs led to reduced cytotoxicity and inflammatory response due to reduced CNT availability. Further investigations showed that both SWCNTs and MWCNTs caused decreased cell viability and impaired phagocytic function. SWCNTs were also shown to inhibit human embryonic kidney cells by inducing apoptosis and decreasing cellular adhesion. However, in a vast majority of these studies, there is inadequate information with regard to the CNT type, synthesis route, purification steps, handling steps, impurity levels, etc. The lack of systematic characterization of the material tested continues to throw the toxicological impact of CNTs into doubt, and it remains a possibility that well-dispersed, highly purified, catalyst-free, surface-functionalized CNTs could yet prove to be a viable material for biomedical applications such as vectors for drug and gene delivery.

In contrast to *in vivo* toxicity of CNTs, *in vitro* studies have shown that CNT-coated substrates and 3D scaffolds offer positive indications in interactions with a variety of cells. For instance, both pristine and functionalized CNT surfaces have a positive impact on neuronal growth and demonstrated no cytotoxicity towards neuronal cells. This is of particular significance owing to the electrical conductivity of CNTs and the potential to electrically stimulate or interface with neurons. In addition to pristine and functionalized CNTs, composite materials such as PU and CNT composites have been evaluated as scaffolds, allowing for variation of CNT loading. For applications as prosthetic and implant materials, it has been shown that CNT composites promote adhesion of chondrocytes and osteoblasts and increased proliferation of osteoblasts, chondrocytes, fibroblasts and smooth muscle cells. No cytotoxicity was reported towards any of these cells.

12.4.2   Biocompatibility of graphene materials

There have been multiple conflicting reports about the biocompatibility and antimicrobial activity of GO. Lately, it is perhaps becoming clearer that purified and well-prepared GO is not antibacterial. In one experiment [23] (Figure 12.6), Luria–Bertani (LB) nutrient broth containing 25 μg mL^{-1} GO was inoculated with *Escherichia coli* bacterial cells to a

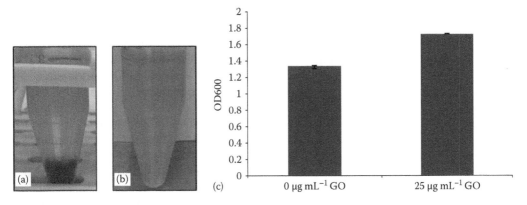

FIGURE 12.6

Bacterial proliferation in the presence of colloidal graphene oxide. Photographs showing bacterial growth in test tubes containing 5 mL of LB broth with (a) 25 µg mL^{-1} and (b) 0 µg mL^{-1} graphene oxide. (c) Graph showing bacterial growth levels in the supernatant of samples containing and lacking GO. (Reprinted with permission from Ruiz, O.N. et al., *ACS Nano*, 5, 8100. Copyright 2011 American Chemical Society.)

concentration of 0.03 OD and incubated for 16 h at 37 °C. The results showed that the GO-containing samples achieved an average absorbance of 1.7 in 16 h of incubation whilst the bacteria growing in LB broth only achieved an absorbance of 1.3. These results and others show that bacteria grew faster and to a higher optical density than in cultures without GO, indicating that GO is not a bactericidal or bacteriostatic material, but instead a general growth enhancer that acts as a scaffold for cell surface attachment and proliferation.

Numerous studies have explored the biocompatibility and toxicology of GO with various animal and human cell types, and in general, GO has proven biocompatible and non-toxic at moderate dosages. Zhang et al. evaluated the biocompatibility of GO by monitoring the effect of concentration of GO and incubation time on the morphological changes of human erythrocytes [24]. It was observed that that almost all the erythrocyte membranes were kept integrated when they incubated with PBS for up to 4 h. Although GO flakes were adhered to the surface of red blood cells, GO suspension showed little effect on the erythrocyte morphology and membrane integrity at the dosage of 10 µg mL^{-1} for 1 and 4 h. However, a part of erythrocyte membranes were ruptured and ghost cells were observed when erythrocytes exposed to 80 µg mL^{-1} of GO for 4 h. Since GO is biocompatible with blood cells, this will pave the way for further development of GO for targeted drug delivery and other biomedical applications.

Whilst graphene is intrinsically non-dispersible in water, various modifications and functionalizations can be undertaken to make it dispersible and stable in water and other aqueous physiological media. For applications in drug delivery, very small flakes of graphene known as nano-graphene sheets (NGS) or graphene quantum dots are used. The toxicity of these flakes was evaluated by Chong et al. [25]. The nanographene samples were obtained through oxidative cutting from graphite. The size of the NGS particles can be seen in the AFM image to be about 10s of nanometres. The NGS was subsequently functionalized with polyethylene glycol (PEG) to render it water-dispersible and bio-compatible. WST-1 assay, which reveals the mitochondrial function of cells demonstrated no toxicity of NGS-PEG to human epithelial cells. Immunochemistry experiment for apoptosis and necrosis further demonstrated negligible effects of NGS at the cellular level. The NGS did not induce any apoptosis or necrosis of human epithelial cells even at the concentration of 160 mg mL^{-1} of NGS, and the apoptosis levels are irrelevant to the dose of the NGS. Overall, the NGS-PEG exhibits no apparent *in vitro* and *in vivo* toxicity. Since these NGS are intrinsically fluorescent, they could be used for bio-imaging and sensing platform, and drug delivery systems.

12.4.3   Biodistribution of graphene materials

The next consideration is the biodistribution of graphene particles, either as GO or otherwise functionalized and water-dispersible graphene, from various routes of uptake. Liu et al. studied the biodistribution of GO flakes of different sizes at different doses in healthy adult male mice by intravenous administration [26]. Two sizes of graphene flakes, small (s-) and large (l-), were labelled with ^{125}I, which allowed a quantitative determination of GO in tissues. After intravenous injection, s-GO mainly accumulated in liver, with a few particles in the lungs and spleen. However, s-GO was scarcely found in other organs, including the brain, stomach, muscle, bone and intestine. The uptake reached the maximum at 5 min post-exposure and the uptake of s-GO in the liver and lungs decreased dramatically following that. For l-GO, instead of the liver, the lungs became the main accumulating organs. Other organs only capture a small amount of l-GO. As time elapsed, the amount of l-GO in lungs decreased. Transmission electron microscopy (TEM) was used to visualise s-GO in the liver and l-GO in the lungs. The nanosized particles and aggregates could be easily distinguished intracellularly in the liver at 10 min post-exposure to s-GO. More GO accumulated in phagocytic cells, whilst a few particles were found in erythrocytes. GO was seldom observed in other cells. On the other hand, at 10 min post-exposure to l-GO, particles larger than 1 μm could be observed in the cell gap in lungs, rather than inside the cells. Both GO samples were rapidly cleared from the blood circulation, however, they accumulated to a significant extent in reticuloendothelial system (RES) organs and showed slow clearance from the mouse body. The less dispersed l-GO formed larger GO–protein complexes and was filtered by the pulmonary capillary vessels.

Yang et al. looked at the long-term *in vivo* biodistribution of ^{125}I-labelled NGS functionalised with PEG [27]. Atomic force microscope (AFM) images showed that NGS-PEG were very small sheets with a size range of ca. 10–30 nm. It was found that NGS-PEG distributed in many different organs at 1 h post-injection, but mainly accumulated in the RES such as liver and spleen at later time points. Brown-black spots which could be clearly differentiated from the blue-stained cell nuclei were noted in the liver of mice 3 days after injection of graphene. Far fewer black spots in the liver were observed 20 days later. However, the black spot numbers in liver slices decreased with time. Substantial bone uptake of NGS was also noted at early time points post-injection, likely owing to the macrophage uptake of nanomaterials in the bone marrow. The kidney and intestine uptake of NGS could be associated with possible renal and faecal excretion, respectively. Ultimately, no obvious toxicity of NGS-PEG at the dose of 20 mg kg^{-1} to female mice was observed in blood biochemistry, haematology and histology analysis.

12.4.4 Graphene materials as scaffolds for tissue engineering

GO thin films form a biocompatible artificial surface for the growth of mammalian cells, for example fibroblast cells as demonstrated by Ryoo et al. [28]. This work compares glass coverslips with no coating, GO coating and rGO coating by proliferation assay, cell shape analysis, focal adhesion study and quantitative measurements of cell adhesion-related gene expression levels. Fluorescence microscopy revealed that most of the cells plated on each substrate were alive, and the viabilities were similar to that on sterile glass, which was used as a control. The proliferation rates, which were calculated by counting the number of live cells per unit area on each substrate, were very similar among the different substrates. However, compared to bare glass, stronger cell–substrate interactions with graphene lead to the development of more focal adhesion sites (Figure 12.7a). The focal adhesions of cells cultured on glass were generally larger and fewer per cell than those of cells cultured on the graphene-coated substrates. This and other such studies suggest that these graphene holds high potential for bioapplications showing high biocompatibility, especially as surface coating materials for implants, without inducing notable deleterious effects.

Induced pluripotent stem cells (iPSCs) hold great promise as a cell source for regenerative medicine yet its culture, maintenance of pluripotency and induction of differentiation remain challenging. GO and rGO have been shown to support the iPSCs cultures and allow for spontaneous differentiation [29]. iPSCs in 2D films or in suspension cultures

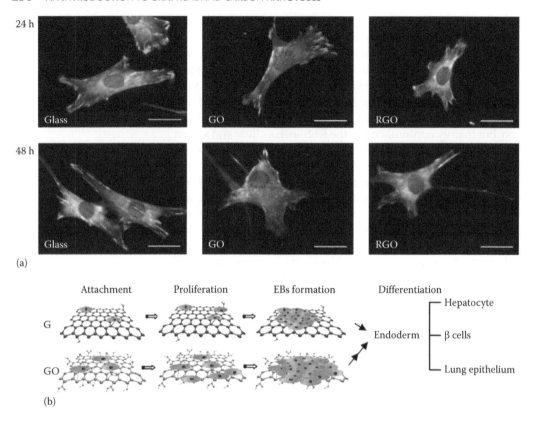

FIGURE 12.7
(a) Comparison of cell adhesion patterns of NIH-2T3 fibroblasts on glass, GO and rGO shows significantly higher number of focal adhesions per cell on GO. (b) Schematic use of graphene and GO as a platform for iPSCs culture and differentiation. ([a]: Reprinted with permission from Ryoo, S.R., Kim, Y.K., Kim, M.H. and Min, D.H., *ACS Nano*, 4, 6587. Copyright 2010 American Chemical Society; [b]: Reprinted from *Biomaterials*, 33, Chen, G.Y., Pang, D.W.-P., Hwang, S.-M., Tuan, H.-Y. and Hu, Y.-C., 418, Copyright 2012, with permission from Elsevier.)

spontaneously differentiate *in vitro* and form 3D aggregates known as embryoid bodies (EBs) which encompass cells of endodermal, mesodermal and ectodermal lineages. As differentiation continues, a variety of cell types are developed within the EBs environment. EBs on the GO surface compared to graphene and glass surfaces were strikingly larger in size (Figure 12.7b), which was concomitant with the higher initial cell density attached to the GO surface. This and other experiments demonstrated that graphene and GO were biocompatible with iPSCs and supported iPSCs attachment and proliferation. Graphene and GO also influence iPSCs differentiation. Differentiation along the endodermal pathway was mitigated on graphene but was promoted on GO; both substrates exhibited similar degrees of ectodermal and mesodermal differentiation. Altogether, it can be seen that rGO and GO could serve as platforms for iPSCs culture and differentiation. Both rGO and GO surfaces supported the iPSCs culture, allowed for spontaneous differentiation, but led to different cell attachment, proliferation, EBs formation and differentiation characteristics.

Neural stem cell (NSC)-based therapy provides a promising approach for neural regeneration. For the success of NSC clinical application, a scaffold is required to provide 3D cell growth microenvironments and appropriate synergistic cell guidance cues. Graphene foam, similar to a graphene aerogel, has a 3D porous structure and can serve as a novel scaffold for NSCs [30]. A 3D porous structure is different from a 2D stem cell culture system, since the porous graphene foam could provide 3D micro-environment in which cells are able to resemble their *in vivo* counterparts. Also, because neural cells are electro-active and electrical stimulation can affect NSC behaviours, it is promising to develop a robust graphene-based 3D bioelectronic interface to regulate NSC behaviours taking advantage of the conductive properties of graphene.

12.4.5   Graphene materials for drug delivery and gene transfection

In addition to being biocompatible as we have seen, modified forms of graphene can also be used as a vector for drug delivery. Often, the graphene used for drug delivery is what is NGS, with flake sizes of less than 100 nm, small enough to enter individual cells. Other approaches include graphene/polymer composite systems for stimulated drug delivery, as we will see next.

Weaver et al. have shown an electrically controlled drug delivery nanocomposite composed of GO deposited inside a conducting polymer scaffold, which in this case is poly(pyrrole) or PPy [31]. The nanocomposite is loaded with an anti-inflammatory molecule, dexamethasone or DEX, and exhibits favourable electrical properties. In response to voltage stimulation, the nanocomposite releases drug with a linear release profile and a dosage that can be adjusted by altering the magnitude of stimulation. Graphene can also be used as a non-toxic nano vehicle for efficient gene transfection [32].

12.4.6   Graphene materials for cancer therapy

We have previously encountered NGS with PEG functionalization, and it has been shown that this material exhibits excellent tumour uptake and we can utilise the strong optical absorbance of NGS in the near-infrared (NIR) region for *in vivo* photothermal therapy, achieving ultra-efficient tumour ablation after intravenous administration of NGS and low-power near infra-red laser irradiation on the tumour [33,34].

Clinical use of various potent pharmacophoric molecules, many of which are aromatic, is often hampered by their poor solubility in water. One example is Tamoxifen Citrate (TmC), a breast cancer drug, which has only a maximum solubility of 0.5 mg mL^{-1} in water at 37 °C. The hydrophobic nature of the tamoxifen molecule restricts its water solubility and therefore severely limits its administration. NGS exhibit a strong non-covalent association with hydrophobic aromatic drugs via physical adsorption. When the graphene is in turn stabilized in water using cationic amphiphiles such as PY$^+$-Chol, they can be used to effectively solubilize anticancer drugs in water. Figure 12.8 shows schematics of functionalised NGS carrying polyethylene glycol and tamoxifen citrate.

Finally, tumour-initiating cells (TICs), also known as cancer stem cells (CSCs), are difficult to eradicate with conventional approaches to cancer treatment, such as chemotherapy and radiation. As a consequence, the survival of residual CSCs is thought to drive the onset of tumour recurrence, distant metastasis and drug resistance, which is a significant clinical problem for the effective treatment of cancer. CSCs proliferate and form 3D spheroid-like structures, containing CSCs and progenitor cells, which are known as "tumour-spheres" or "onco-spheres". It has been shown that graphene oxide (of both big and small flake sizes) can be used to selectively inhibit the proliferative expansion of cancer stem cells [35], across six independent cancer cell lines, across multiple tumour types (breast, ovarian, prostate, lung and pancreatic cancer, as well glioblastoma (brain cancer)).

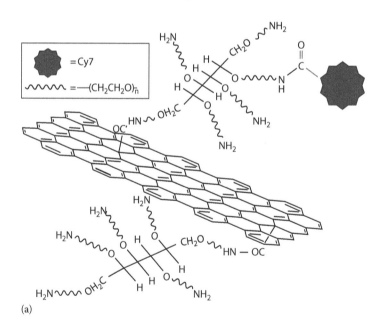

(a)

FIGURE 12.8

Schematics of nanographene sheets with (a) polyethylene glycol functionalization for photothermal cancer therapy. ([a]: Reprinted with permission from Yang, K., Zhang, S., Zhang, G., Sun, X., Lee, S.-T. and Liu, Z., *Nano Lett.*, 10, 3318. Copyright 2010 American Chemical Society.) *(Continued)*

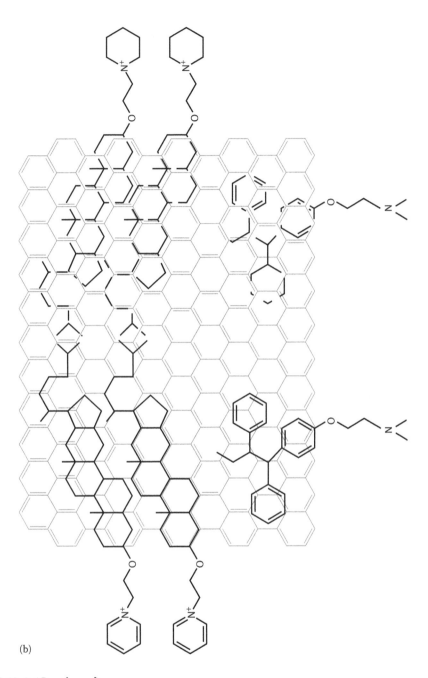

(b)

FIGURE 12.8 (*Continued*)
Schematics of nanographene sheets with (b) Tamoxifen Citrate functionalization for cancer chemotherapy. ([b]: Reprinted from *Biomaterials*, 33, Chen, G.Y., Pang, D.W.-P., Hwang, S.-M., Tuan, H.-Y. and Hu, Y.-C., 418, Copyright 2012 with permission from Elsevier.)

12.5   Carbon nanotubes and graphene for energy storage

12.5.1   Graphene-based supercapacitors

For capacitive energy storage, two types of systems can be used, either an electrical double-layer capacitor (EDLC) or a pseudocapacitor. Pure graphene electrodes can be used for EDLCs, whilst polyannaline is a material of choice for pseudocapacitors. In an EDLC, such as with graphene electrodes, the charge is stored electrostatically in a Helmholtz double layer at the electrode–electrolyte interface. There is no electrode–electrolyte charge transfer. In the case of pseudocapacitors, charge is stored electrochemically or Faradaically, where charge is reversibly transferred between the electrode and electrolyte through redox reactions. EDLCs possess good electrical conductivity and long cycle life, but relatively low specific capacitance, whilst pseudocapacitors have high specific capacitance but poor stability and slow charge–discharge rates. An emerging class of supercapacitors called hybrid supercapacitors utilize both ways of energy storage, and electrodes for such hybrid supercaps could be made of a graphene-PANI layered composite material.

The construction of a graphene-based supercapacitor [36] is shown in Figure 12.9a. The supercapacitor consists of two graphene layers, separated by a porous membrane that is soaked in electrolyte. In the case of this particular graphene supercapacitor the electrodes were made of rGO, mixed with 10 wt.% polytetrafluoroethylene (PTFE) binder. The two rGO electrodes were separated by a thin PP film in 30 wt.% KOH

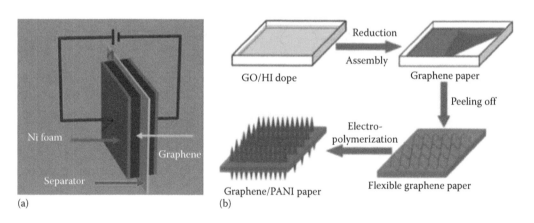

(a)   (b)

FIGURE 12.9

(a) Schematics of a supercapacitor with graphene electrodes. (b) Schematic showing steps in the preparation of a PANI/graphene composite electrode for a hybrid supercapacitor. ([a]: Reprinted with permission from Wang, Y. et al., *J. Phys. Chem. C*, 113, 13103. Copyright 2009 American Chemical Society; [b]: Cong, H.-P., Ren, X.-C., Wang, P. and Yu, S.-H, *Energy Environ. Sci.*, 6, 1185, 2013. Reproduced by permission of The Royal Society of Chemistry.)

aqueous electrolyte solution. The graphene electrodes were pressed on nickel foam current electrodes and were sandwiched in a stainless steel cell with a pressure to form a sealed "coin-cell" type of supercapacitor.

The specific capacitance of this graphene supercapacitor was 205 F g^{-1}, the energy density was 28.5 Wh kg^{-1} and power density was 10 kW kg^{-1}. The high value of the power density is well suited for surge-power delivery applications. The specific capacitance only decreases by 10% after 1200 cycles.

To produce a hybrid supercapacitor [37], rGO paper was produced by reduction of GO using HI and subsequent washing to remove excess iodine. The composite hybrid capacitor electrodes were fabricated by electrochemical polymerization of aniline monomers on the graphene paper to obtain a flexible graphene-PANI paper with a green colour (Figure 12.9b). The PANI grows in the form of nanorods on the two outer graphene paper surfaces but also on the inner surfaces between the graphene sheets, resulting in a 3D laminated composite of graphene and PANI layers.

The specific capacitance of the graphene-PANI electrode was greater than both graphene paper and PANI thin-film which translates to increased energy storage capacity in the hybrid supercapacitor compared to a purely electric double layer supercapacitor or a pseudocapacitor. Also, 64% of the specific capacitance was maintained when the current density went from 1 to 10 A cm^{-1} compared to 56% for pure PANI which leads to faster charge/discharge rates than either a supercapacitor or pseudocapacitor. And finally, the hybrid electrode retained 82% specific capacitance after 1000 cycles compared to 52% for pure PANI. This translates to a longer lifespan.

12.5.2   Carbon nanotube-based lithium-ion batteries

Lithium-ion batteries realise the store and release of energy by the transfer of lithium ions between the anode and cathode. In conventional lithium-ion cells, the cathode is a Li-metal-oxide and the anode is made of graphite. The lithium ions intercalate the layers of graphite reversibly to store and discharge energy. The theoretical limit is one lithium ion for every six carbon atoms, corresponding to a theoretical capacity of 372 mAh g^{-1}. It has been shown that a pure SWCNT electrode can exceed this Li storage limit due to its morphology compared to planar stacked graphite. MWCNTs have been shown to store large amount of lithium ions (952 mAh g^{-1}) [38], whilst the reversible capacity was 447 mAh g^{-1}. Changes such as surface functionalization, reducing the length and nitrogen doping can all increase this capacity even further. These modifications, however, makes the half-cell potential of the SWCNT electrode and consequently the battery's voltage unreliable and unstable, which is a challenge to be overcome. In another design, SWCNTs can be added to silicon or

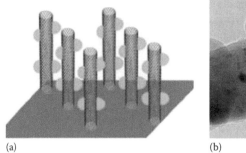
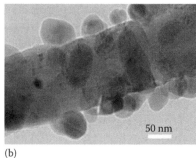

(a)   (b)

FIGURE 12.10
(a) Schematic and (b) scanning electron micrograph of silicon nanoclusters supported on CNT arrays for use as electrodes in Li-ion batteries. (Reprinted with permission from Wang, W. and Kumta, P.N., *ACS Nano*, 4, 2233. Copyright 2010 American Chemical Society.)

metal-oxide anodes to increase the electrical conductivity of the electrode as well as to accommodate expansion during cycling. SWCNT-reinforced silicon electrodes [39] (Figure 12.10) have achieved charge storage capacities in excess of 3000 mAh g^{-1} [40]. On the other side, CNTs can be incorporated into the lithium-metal-oxide cathode to increase the electrode conductivity, cycling stability and charge storage capacity. These strategies have been commercially applied by a number of lithium-ion battery manufacturers and several tens of tonnes of SWCNTs have been consumed as conducting additives for Li-ion batteries.

12.6   Conclusion

Both graphene and carbon nanotubes have a broad range of potential applications, and judging purely by their intrinsic nanoscale properties, they can be expected to significantly outperform existing technologies. In addition to applications discussed here, these materials are also being explored as membranes for filtration and desalination, barrier coatings such as for corrosion protection, conductive inks for printed electronics, scaffolds for catalysis, field-emission electron sources, micro-electro-mechanical systems such as mass sensors, etc. The applications discussed here are what might be considered closer to commercialization. Even in these cases, as we have seen, significant challenges remain in scale-up, manufacturability and quality control. Nonetheless, applications such as CNTs in batteries and aerospace composites have already achieved commercial viability, and it is anticipated that a number of other applications will become commercially viable very soon. It is therefore not unreasonable to assume that we will all be encountering graphene and carbon nanotubes in our everyday life in the near future.

References

1. J Liu and MC Hersam, *MRS Bull.* **35**, 315 (2010).

2. X Tu, S Manohar, A Jagota and M Zheng, *Nature* **460**, 250 (2009).

3. AA Green and MC Hersam, *Mater. Today* **10**, 59 (2007).

4. T Tanaka et al., *Nano Lett.* **9**, 1497 (2009).

5. A Vijayaraghavan, *Phys. Stat. Sol. (b)* **250**, 2505 (2013).

6. M Steiner et al., *Appl. Phys. Lett.* **101**, 053123 (2012).

7. Z Chen, Y-M Lin, MJ Rooks and P Avouris, *Physica E* **40**, 228 (2007).

8. MY Han, B Özyilmaz, Y Zhang and P Kim, *Phys. Rev. Lett.* **98**, 206805 (2007).

9. LA Ponomarenko et al., *Science* **320**, 356 (2008).

10. C Stampfer, E Schurtenberger, F Molitor, J Güttinger, T Ihn and K Ensslin, *Nano Lett.* **8**, 2378 (2008).

11. F Molitor, S Droscher, J Guttinger, A Jacobsen, C Stampfer, T Ihn and K Ensslin, *Appl. Phys. Lett.* **94**, 222107 (2009).

12. N Tombros, C Jozsa, M Popinciuc, HT Jonkman and BJV Wees, *Nature* **448**, 571 (2007).

13. CR Dean et al., *Nat. Nanotechnol.* **5**, 722 (2010).

14. L Britnell et al., *Science* **335**, 947 (2012).

15. SJ Haigh et al., *Nat. Mater.* **11**, 764 (2012).

16. H Kim, Y Miura and CW MacOsko, *Chem. Mater.* **22**, 3441 (2010).

17. AS Wajid, HS Tanvir Ahmed, S Das, F Irin, AF Jankowski and MJ Green, *Macromol. Mater. Eng.* **298**, 339 (2013).

18. D Wang, X Zhang, JW Zha, J Zhao, ZM Dang and GH Hu, *Polymer* **54**, 1916 (2013).

19. K Hu, MK Gupta, DD Kulkarni and VV Tsukruk, *Adv. Mater.* **25**, 2301 (2013).

20. Q Zhang, J-Q Huang, W-Z Qian, Y-Y Zhang and F Wei, *Small* **9**, 1237 (2013).

21. SK Smart et al., *Carbon* **44**, 1034 (2006).

22. L Lacerda, A Bianco, M Prato and K Kostarelos, *Adv. Drug Deliv. Rev.* **58**, 1460 (2006).

23. ON Ruiz et al., *ACS Nano* **5**, 8100 (2011).

24. X Zhang, J Yin, C Peng, W Hu, Z Zhu, W Li, C Fan and Q Huang, *Carbon* **49**, 986 (2011).

25. Y Chong et al., *Biomaterials* **35**, 5041 (2014).

26. JH Liu, ST Yang, H Wang, Y Chang, A Cao, Y Liu, *Nanomedicine* **7**, 1801 (2012).

27. K Yang, J Wan, S Zhang, Y Zhang, ST Lee, Z Liu, *ACS Nano* **5**, 516 (2011).

28. SR Ryoo, YK Kim, MH Kim and DH Min, *ACS Nano* **4**, 6587 (2010).

29. GY Chen, DW-P Pang, S-M Hwang, H-Y Tuan and Y-C Hu, *Biomaterials* **33**, 418 (2012).

30. N Li et al., *Sci. Rep.* **3**, 1604 (2013).

31. CL Weaver, JM LaRosa, X Luo and XT Cui, *ACS Nano* **8**, 1834 (2014).

32. L Feng, S Zhang and Z Liu, *Nanoscale* **3**, 1252 (2011).

33. K Yang, S Zhang, G Zhang, X Sun, S-T Lee and Z Liu, *Nano Lett.* **10**, 3318 (2010).

34. SK Misra, P Kondaiah, S Bhattacharya and CNR Rao, *Small* **8**, 131 (2012).

35. M Fiorillo et al., *Oncotarget* **6**, 3553 (2015).

36. Y Wang et al., *J. Phys. Chem. C* **113**, 13103 (2009).

37. H-P Cong, X-C Ren, P Wang and S-H Yu, *Energy Environ. Sci.* **6**, 1185 (2013).

38. E Frackowiak, S Gautier, H Gaucher, S Bonnamy and F Beguin, *Carbon* **37**, 61 (1999).

39. W Wang and PN Kumta, *ACS Nano* **4**, 2233 (2010).

40. A Gohier et al., *Adv. Mater.* **24**, 2592 (2012).

41. A Vijayaraghavan, S Blatt, D Weissenberger, M Oron-Carl, F Hennrich, D Gerthsen, H Hahn, and R Krupke, Ultra-large-scale directed assembly of single-walled carbon nanotube devices. *Nano letters*. 2007 Jun 13;7(6):1556–60.

Appendix A: Raman Scattering in Non-Molecular Solids

Raman scattering has played an important role in the characterization of all carbon-based materials and a short introduction to Raman scattering in non-molecular solids is therefore presented here. The interested reader is referred to texts treating the subject in more detail ([4,19] of Chapter 5 and [1,2]).

A.1 Energy, momentum and wavenumber in Raman scattering

Our primary aim in this appendix is to be able to relate the observed Raman spectrum of a sample to the phonon dispersion relation and density of states. This task is accomplished mainly by considering the conservation of energy and momentum in Raman scattering, and by understanding the role of sample polarizability in causing Raman scattering.

Raman scattering in non-molecular solids is the inelastic scattering of photons by a sample, when phonons in the sample are either created or destroyed. Here we shall primarily discuss first-order Raman scattering in which a single phonon is created or a single phonon is destroyed.

The process in which a phonon is created (and the photon therefore loses some energy in the process) is called Stokes Raman scattering and the process in which a phonon is destroyed (and the photon gains some energy in the process) is called anti-Stokes Raman scattering.* These processes are frequently visualized as involving the excitation of the system to a virtual energy level (Figure A.1) despite the fact that in general terms there is no physical rationale for the concept of the virtual energy level. When studying the allotropes of carbon, however, the concept of the virtual energy level in Raman scattering is relevant because around half of the Raman peaks observed from carbon allotropes originate from resonant Raman scattering processes in which an electron is excited to the conduction band (so there is a real energy level rather than a virtual energy level).

Utilizing Figure A.1, we can now write down the conditions for the conservation of energy (Equation A.1) and momentum (Equation A.2) in a first-order Raman scattering process:

$$\hbar\omega_i = \hbar\omega_s \pm \hbar\omega_p \tag{A.1}$$

$$\hbar\mathbf{k}_i = \hbar\mathbf{k}_s \pm \hbar\mathbf{k}_p \tag{A.2}$$

where
- ω_i and ω_s are the frequencies of the incident and scattered photon
- ω_p is the frequency of the phonon created (destroyed) in the Stokes (anti-Stokes) Raman scattering process
- $\mathbf{k}_i$, $\mathbf{k}_s$ and $\mathbf{k}_p$ are the wavevectors of the incident and scattered photon and the phonon respectively

---

* The terms "Stokes" and "anti-Stokes" scattering apply to all Raman scattering, not just to the simplest case of non-resonant first-order Raman scattering described in this section.

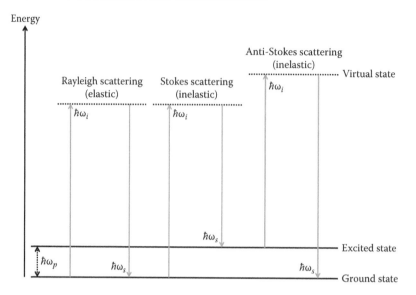

FIGURE A.1

Schematic energy level diagram of first-order Stokes and anti-Stokes Raman scattering processes. The incident laser photon has energy $\hbar\omega_i$ and the phonon excited (annihilated) in Stokes (anti-Stokes) scattering has energy $\hbar\omega_p$. The scattered photon has energy $\hbar\omega_s$.

In a typical crystalline solid, the real space lattice constants are of the order of 5 Å. Wavevectors for phonons in the first Brillouin zone therefore span from $k_p = 0$ to $k_p \sim 10^{10}$ m^{-1}. In contrast, Raman scattering processes are typically excited with visible laser photons (for instance green excitation with $\lambda \approx 500$ nm). The wavevector of the incoming photon is therefore $k_i \sim 10^7$ m^{-1}. Therefore, $k_i \ll k_p$ for all phonons except those at the very centre of the Brillouin zone. The conservation of momentum (Equation A.2) therefore dictates that in a conventional first-order Raman scattering process, it is only possible to excite or destroy phonons very close to the centre of the first Brillouin zone ($k_p \approx 0$) [1,2]. This is the fundamental selection rule in Raman scattering.*

A.1.1   Units of measurement in Raman scattering

When we perform a Raman scattering experiment, we measure the energy of the Raman-active phonon(s) in the sample under investigation. It is, however, conventional to use units of cm^{-1}. So, for instance, the most intense peak in the Raman spectrum of silicon is at $E = 65$ meV, which is conventionally recorded as $\bar{v} = 520$ cm^{-1} using the conversion $E = hc\bar{v}$.

Arguably this is confusing since we are using units of cm^{-1} to describe the energy of the photon, and not the wavevector (which has to be close to zero for a phonon excited in a first-order non-resonant Raman scattering process). There is, however, logic behind this choice of units. When Raman scattering is performed experimentally, the Raman scattered photons are diffracted at a diffraction grating and the raw data recorded is the angle through which

---

* Readers may have studied Raman scattering from molecules in which vibrational or rotational energy levels of the molecule are excited. The reason such treatments of Raman scattering do not include this selection rule is because for any vibrational or rotational energy of a molecule, $q_p = 0$. The excitation cannot have momentum if it is not propagating through a sample – whilst a phonon can propagate through a non-molecular solid a vibrational or rotational excitation in a molecule must stay in the same molecule.

they are diffracted. This is directly determined by the relation between the wavelength (or wavenumber) of the photon, and the line spacing on the diffraction grating. Converting into units of energy using $E = hc\bar{v}$ requires knowledge of the speed of light and Planck's constant. Whilst these values are nowadays established to very many decimal places, when Raman scattering was first developed as an analytical tool (the 1930s), there was still a possibility that the accepted values of these constants could change. It was therefore appropriate for scientists to report their results in units of cm^{-1} so that the values reported were independent of any assumptions about the value of the speed of light and Planck's constant.

A.2 Polarizability in Raman scattering: Optical and acoustic phonons

Thus far, we have understood that we can only excite zone-centre phonons in a Raman scattering process. There are, however, additional restrictions on which phonons can be excited, and we can understand why this is the case by studying the change in electric susceptibility of a sample due to the excitation of phonons ([4] of Chapter 5). We can begin by considering the illumination of the sample with a sinusoidal electromagnetic wave having wavenumber k_i and frequency ω_i (the excitation laser in a Raman spectroscopy experiment). The electric field amplitude experienced by the sample is given by

$$E(r,t) = E_o \cos(k_i \cdot r - \omega_i t)$$

This field polarizes the sample, with polarization given by

$$P(r,t) = \chi\varepsilon_0 E(r,t) = \chi\varepsilon_0 E_o \cos(k_i \cdot r - \omega_i t) \tag{A.3}$$

where χ is the dimensionless electric susceptibility of the sample. For simplicity, let us assume that the sample is isotropic and that χ is hence a scalar quantity. In the absence of phonons, χ is a constant but in the presence of phonon(s) the value of χ oscillates. Let us therefore define the oscillating displacement D of an atom when a phonon is excited:

$$D(r,t) = A\cos(k_p \cdot r - \omega_0 t) \tag{A.4}$$

The value of χ is determined by the crystal structure of the sample. Let us assume that the vibrations of the atoms in the sample caused by the phonon are small enough not to cause significant changes to the crystal structure. We can then conclude that any change in χ caused by the presence of the phonon will also be small and represent the oscillation in χ about its value in the absence of phonons (χ_0) using the first term in a Taylor series:

$$\chi(D) = \chi_0 + |\nabla_A\chi|_{A=0}|D(r,t)| \tag{A.5}$$

The polarization (Equation A.3) in the presence of the phonon is hence

$$P(r,t) = \varepsilon_0\left[\chi_0 + |\nabla_A\chi|_{A=0}|D(r,t)|\right]E_o \cos(k_i.r - \omega_i t) \tag{A.6}$$

We now divide the polarization into two components, the static polarization in the absence of the phonon P_0 and the dynamic polarization induced by the phonon P_{dyn}. These are represented by the two terms in Equation A.6. We can therefore (substituting for $D(r,t)$ in Equation A.6) write an expression for P_{dyn}:

$$P(r,t) = P_0 + P_{dyn} \tag{A.7}$$

$$P_0 = \varepsilon_0 \chi_0 E_0 \cos\left(k_i \cdot r - \omega_i t\right) \tag{A.8}$$

$$P_{dyn} = \varepsilon_0 \left|\nabla_A \chi\right|_{A=0} \left|A\right| E_o \cos\left(k_p \cdot r - \omega_0 t\right) \cos\left(k_i \cdot r - \omega_i t\right) \tag{A.9}$$

This can be rearranged to give*

$$P_{dyn} = \frac{1}{2}\varepsilon_0 \left|\nabla_A \chi\right|_{A=0} \left|A\right| E_o \left[\cos\left[\left(k_i + k_p\right)\cdot r - \left(\omega_i + \omega_0\right)t\right] + \cos\left[\left(k_i - k_p\right)\cdot r - \left(\omega_i - \omega_0\right)t\right]\right] \tag{A.10}$$

Referring to the two terms comprising Equations A.7 through A.10, we can see that in the absence of any effects due to phonons the sample radiates radiation at the same frequency ω_i (and hence energy) as the incoming radiation (Equation A.8). However, due to the change in electric susceptibility in the presence of phonons the sample also radiates at frequencies $\omega_i + \omega_0$ and $\omega_i - \omega_0$ (Equation A.10). This is the Stokes and anti-Stokes Raman scattering as discussed earlier.

No Raman scattering occurs ($P_{dyn} = 0$) unless $\left|\nabla_A \chi\right|_{A=0} \neq 0$. Physically, this means that for a phonon to be Raman-active $k_p \approx 0$ and the electric susceptibility must change upon creation of the phonon even in the limit that the displacement $A \rightarrow 0$. Note this is a slightly more stringent condition than that frequently quoted in the literature (that the electric susceptibility must change upon creation of the phonon i.e. it is an optical phonon). Gallium nitride in the hexagonal (wurtzite) structure provides a good example to illustrate this; not all $k_p \approx 0$ optical phonons appear in the Raman spectrum [3,4]. This is also the case for graphene and graphite ([5] of Chapter 6).

We refer to phonons which cause the electric susceptibility to change as optical phonons because they can (usually) be excited at the zone-centre by electromagnetic radiation. These tend to be the higher energy phonons and involve the atoms vibrating out of phase. The branches of the phonon dispersion relation emanating from $E = 0$ at the zone-centre, on the other hand, are called acoustic phonons. They involve atoms vibrating in phase and in the limit $E \rightarrow 0$, $k \rightarrow 0$ these are the phonons responsible for the propagation of sound waves through a material.

In addition, we can note that since metals have zero susceptibility it is generally not possible to observe a Raman spectrum from metals; there are some exceptions in metals with the hexagonal close-packed structure [5].

A.3   Raman scattering in defective and amorphous materials

The phonon dispersion relation of graphene is a good example of the relation between frequency (i.e. energy) and wavevector for phonons in an infinitely large crystal (at least in the xy plane). As a result of the symmetry of the crystal lattice, phonons only exist with very specific relationships between energy and wavevector. Whilst only optical phonons at the centre of the Brillouin zone can be excited using first-order Raman scattering, a wider variety of phonons can often be excited using second-order Raman scattering (for instance, in graphite and graphene), it is possible to excite phonons at the edge of the Brillouin zone using Brillouin scattering ([4] of Chapter 5), and all phonons can be excited using inelastic neutron scattering ([2] of Chapter 1 and [6]). So theoretical predictions of phonon dispersion relations can be verified rigorously.

---

* $\cos\theta + \cos\varphi = 2\cos\left[\frac{1}{2}\left(\theta + \varphi\right)\right]\cos\left[\frac{1}{2}\left(\theta - \varphi\right)\right]$

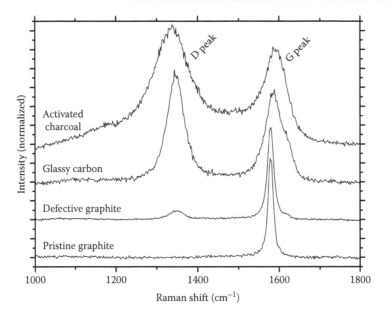

FIGURE A.2
Raman spectra of pristine graphite, defective graphite, glassy carbon and activated charcoal (normalized to the same G peak intensity). The D peak intensity rises as the defect concentration rises, as does the width of both peaks.

On the other hand, in a completely amorphous material there is no longer a specific dispersion relation between the phonon energy and wavevector. The Raman spectrum as a function of phonon energy therefore resembles the entire phonon density of states for the amorphous material [7–9]. If we start with a single infinitely large crystal and gradually divide it up into smaller and smaller crystallites the selection rules as described earlier are gradually relaxed and the Raman spectrum undergoes a gradual change from that expected from the crystalline solid to a representation of the phonon density of states for the amorphous material. Gradually increasing the concentration of defects has the same effect.

Alternatively, one can understand the effect of a small concentration of defects as not affecting the phonon dispersion relation per se. The scattering of electrons by the defects can allow momentum to be conserved when a phonon not at the zone centre is created.

Figure A.2 shows first order Raman spectra of pristine graphite, graphite with defects, glassy carbon (i.e. more defects) and amorphous carbon to illustrate the evolution of the Raman spectra with degree of structural order, as the selection rules are gradually relaxed with the increasing level of disorder.

A.4 Temperature in Raman scattering

Our classical treatment of Raman scattering in section A.2 took no account of the quantization of photons and phonons, and the requirement for conservation of energy specifically with regard to this quantization. This must be accounted for in order to explain the relative intensities of the Stokes and anti-Stokes Raman peaks and their variation as a function of temperature.

In Stokes scattering (Figure A.1), the incident photon loses some energy in the scattering process, creating a phonon with this energy. In anti-Stokes scattering, the incident photon gains some energy in the scattering process through the annihilation of a phonon. Hence anti-Stokes scattering can only take place if there is a phonon already present. The intensity of the anti-Stokes Raman signal for a given phonon hence depends on the number N of phonons present of that specific variety. Since phonons are bosons, this is given (Equation A.11) by the Bose–Einstein distribution multiplied by the density of quantum states $D(E)$ available for the phonons:

$$NdE = \frac{1}{e^{\frac{E}{k_B T}} - 1} \times D(E)dE \tag{A.11}$$

The Maxwell–Boltzmann distribution ($NdE = e^{-(E/k_B T)} \times D(E)dE$) is frequently used as an approximation for the Bose–Einstein distribution in this context, on the assumption that $e^{(E/k_B T)} \gg 1$. In either case, the intensity of the anti-Stokes Raman scattering drops exponentially with increasing phonon energy for higher energy phonons. Significant intensity of anti-Stokes Raman scattering is only observed when there is adequate thermal energy available to excite the necessary phonons ($k_B T \sim E$).

For this reason it is possible to measure temperature by recording the relative intensity of the Stokes and anti-Stokes Raman peaks, provided that adequate calibration is carried out to account for the variation in detection efficiency of the spectrometer as a function of photon energy. Usually this is done by commencing with a measurement at ambient temperature.

Figure A.3 demonstrates the decrease in anti-Stokes Raman scattering intensity as a function of energy by comparing the first-order Raman spectra of diamond and silicon at ambient temperature. Since these materials have the same crystal structure and (sp³) bonding their Raman spectra are identical save for the fact that the peaks occur at significantly lower energy in the case of silicon. The intense first-order Raman peak in diamond due to

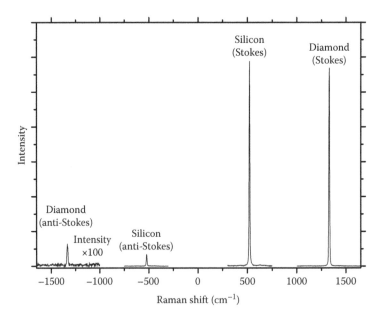

FIGURE A.3

First-order Raman spectra of diamond and silicon, scaled so that the intensities of the Stokes Raman peak are identical. In silicon, the anti-Stokes Raman peak is observed with much greater intensity relative to the Stokes peak.

the zone-centre optical phonon occurs at 1332 cm^{-1} whilst in silicon it occurs at 520 cm^{-1}. Converting 300 K into units of cm^{-1} using $E = (3/2)k_BT = hc\tilde{v}$ gives $\tilde{v} = 312$ cm^{-1}. Hence there is adequate thermal energy available at ambient temperature to excite the phonon responsible for the silicon Raman peak but not the diamond Raman peak. For this reason, the anti-Stokes Raman peak is observed with much greater intensity (relative to the Stokes peak) in the silicon spectrum than in the diamond spectrum.

Generally, the Stokes Raman peaks from a sample are much more intense than the anti-Stokes peaks. Therefore, unless otherwise stated, spectra shown in the literature are the Stokes peaks. These are labelled with a positive wavenumber and the anti-Stokes peaks with a negative wavenumber.

References

1. B Schrader (ed.), *Infrared and Raman Spectroscopy*, Wiley (1995).

2. H Kuzmany, *Solid-State Spectroscopy*, Springer (2009).

3. JA Freitas Jr. and WJ Moore, *Braz. J. Phys.* **28**, 12 (1998).

4. HG Siegle, L Kaczmarkzyk, L Filippidis, A Litvinchuk, C Hoffmann and C Thomsen, *Phys. Rev. B* **55**, 7000 (1997).

5. AA Abrikosov and VM Genkin, *JETP* **65**, 842 (1973).

6. RA Cowley, ADB Woods and G Dolling, *Phys. Rev.* **150**, 487 (1966).

7. R Shuker and RW Gammon, *Phys. Rev. Lett.* **25**, 222 (1970).

8. H Richter, ZP Wang and L Ley, *Solid State Commun.* **39**, 625 (1981).

9. Z Iqbal and S Vepřek, *J. Phys. C.: Solid State Phys.* **15**, 377 (1982).

Appendix B: Additional Notes on the Application of Tight-Binding Theory to Graphene

In Chapter 3, a simple application of tight-binding theory to the π-electrons in graphene was presented. The theory presented in Chapter 3 was chosen for its brevity, and also because the relation to molecular orbital theory as presented in Chapter 2 is made very clear.

In this appendix, a much more rigorous version of the application of tight-binding theory to graphene is presented, in which the compliance with Bloch's theorem is also made completely clear. Following on from this, we are able to derive the Dirac-like tight-binding Hamiltonian matrix for the electronic dispersion relation close to the K point, which is necessary to understand the relativistic nature of the electrons in graphene and the quantum Hall effect in graphene.

The interested reader is referred to the various tight-binding treatments of the graphene lattice, performed both prior to and since the experimental discovery of graphene ([6–8,12] of Chapter 1 and [2] of Chapter 3), and also to general texts covering tight-binding theory ([3] of Chapter 1 and [1]).

B.1 Bloch's theorem

Any rigorous description of tight-binding theory must begin with a statement of Bloch's theorem. Bloch's theorem states that the solutions to the Schrödinger equation for a periodic potential (such as the solutions for the electron wave functions in a crystalline solid) must be wave functions that consist of a function with the same periodicity as the potential, multiplied by a plane wave. One can apply Bloch's theorem to any particle subjected to a periodic potential. When applying it to electrons in a crystalline solid, one can understand it intuitively as a statement that the electron probability density $|\Phi|^2$ in the crystal must have the same periodicity as the crystal itself.

Mathematically, we can apply Bloch's theorem to electrons in graphene as follows. The potential in which the electrons move has the periodicity given by the lattice vectors (in the case of graphene, these are simply a_1 and a_2 as defined in Equation 1.1) and we are interested in the effect on the electron wave function of a translation T which is any integer multiple of the lattice vectors:

$$T = n_1 a_1 + n_2 a_2$$

Bloch's theorem therefore states that under translation T for any integer n_1 and n_2, the eigenfunction $\Phi(r)$ for the electron moving in the periodic potential due to the graphene lattice must be invariant under the translation T except for a phase factor:

$$\Phi(r + T) = e^{ik \cdot T} \Phi(r) \tag{B.1}$$

Hence, the experimental observable $|\Phi|^2$ will have the same periodicity as the potential. General proofs of Bloch's theorem can be found in standard solid state physics textbooks ([2–5] of Chapter 1).

B.2   Application of tight-binding theory to graphene

To apply tight-binding theory to graphene, we must first write down expressions for the electron wave functions in the graphene crystal which comply with Bloch's theorem. In Chapter 2, we dealt with the two atoms in the primitive unit cell of graphene, and as a result we had an electronic dispersion relation with two different branches. This is, in fact, a general result. n atoms in the unit cell leads to a dispersion relation with n branches.

For the case of the electronic dispersion relation of the π-electrons in graphene, we will write down expressions for the two different normalized Bloch wave functions* for an electron in the graphene lattice using the normalized wave functions for the $\Psi(2p_z)$ orbitals for the A and B atoms in the unit cell:

$$\varphi_{A,B}(k,r) = \frac{1}{\sqrt{N}} \sum_{l=1}^{N} e^{ik \cdot R(A,B)l} \Psi(A,B)_l \tag{B.2}$$

Here, there are N unit cells in the crystal, and two atoms (the A and B atoms) in each unit cell. $\Psi(A,B)_l$ is the $\Psi(2p_z)$ orbital for the A, B atom in the lth unit cell, and $R(A,B)_l$ is the location of this atom.

However, the most general solutions possible are linear combinations of the Bloch functions specified in Equation B.2, which are of course also Bloch functions. We expect two Bloch functions $\Psi_1(k,r)$ and $\Psi_2(k,r)$, determined by the values of four coefficients $C_1(A)$, $C_1(B)$, $C_2(A)$ and $C_2(B)$ (Equation B.3). This leads to two eigenvalues for the energy ($E_1(k)$ and $E_2(k)$) which are determined in the normal manner by applying the Hamiltonian operator $\hat{H}$ for the graphene crystal (Equation B.4):

$$\Psi_i(k,r) = \sum_{A,B} C_i(A,B) \varphi_{A,B}(k,r) \tag{B.3}$$

$$E_i(k) = \frac{\int \Psi_i^*(k,r) \hat{H} \Psi_i(k,r) d\tau}{\int \Psi_i^*(k,r) \Psi_i(k,r) d\tau} = \frac{\langle \Psi_i | H | \Psi_i \rangle}{\langle \Psi_i | \Psi_i \rangle} \tag{B.4}$$

Substituting the expressions from Equations B.3 into B.4[†] gives

$$E_i(k) = \frac{C_i^*(A)C_i(A)H_{AA} + C_i^*(A)C_i(B)H_{AB} + C_i^*(B)C_i(A)H_{BA} + C_i^*(B)C_i(B)H_{BB}}{C_i^*(A)C_i(A)S_{AA} + C_i^*(A)C_i(B)S_{AB} + C_i^*(B)C_i(A)S_{BA} + C_i^*(B)C_i(B)S_{BB}} \tag{B.5}$$

where

$$H_{AA} = \left\langle \varphi_A \middle| \hat{H} \middle| \varphi_A \right\rangle, \quad H_{AB} = \left\langle \varphi_A \middle| \hat{H} \middle| \varphi_B \right\rangle, \quad S_{AA} = \left\langle \varphi_A \middle| \varphi_A \right\rangle, \text{ etc.} \tag{B.6}[‡]$$

We now seek to find the values of the coefficients $C_i(A,B)$ that give the lowest values for the energy eigenvalues. We do so by differentiating equation (B.5) with respect to any one of the coefficients (we choose $C_i^*(A)$) and setting the derivative to zero. Performing the differentiation, we will recall that since $C_i(A)$ is a complex quantity, $C_i(A)$ and $C_i^*(A)$

---

* Wave functions satisfying Bloch's theorem.
† These integrals are over all space. The notation we use on the right is known as bra-ket notation.
‡ Notation has been shortened from $\varphi_{A,B}(k,r)$ to $\varphi_{A,B}$.

can be varied independently. The algebra is simplified and printing costs are reduced considerably if we define new variables such that $E_i(k) = (h/w)$ before differentiating using the quotient rule:

$$\frac{\partial E_i(k)}{\partial C_i^*(A)} = \frac{w \frac{\partial h}{\partial C_i^*(A)} - h \frac{\partial w}{\partial C_i^*(A)}}{w^2} = 0 \tag{B.7}$$

$$\frac{\partial h}{\partial C_i^*(A)} = \frac{h}{w} \frac{\partial w}{\partial C_i^*(A)} = E_i(k) \frac{\partial w}{\partial C_i^*(A)}$$

We therefore obtain

$$C_i(A)H_{AA} + C_i(B)H_{AB} = E_i(k)\big(C_i(A)S_{AA} + C_i(B)S_{AB}\big) \tag{B.8}$$

Performing the differentiation on Equation B.5 with respect to $C_i^*(B)$ instead gives

$$C_i(A)H_{BA} + C_i(B)H_{BB} = E_i(k)\big(C_i(A)S_{BA} + C_i(B)S_{BB}\big) \tag{B.9}$$

If we differentiate with respect to $C_i(A)$ or $C_i(B)$, we obtain equations identical to (B.8) and (B.9). Thus these equations are the only conditions that must be satisfied to minimize the energy eigenvalues.

Equations B.8 and B.9 can be conveniently combined in matrix form as follows:

$$HC_i = E_i(k)SC_i \tag{B.10}$$

where

$$H = \begin{pmatrix} H_{AA} & H_{AB} \\ H_{BA} & H_{BB} \end{pmatrix} \tag{B.11}$$

$$S = \begin{pmatrix} S_{AA} & S_{AB} \\ S_{BA} & S_{BB} \end{pmatrix} \tag{B.12}$$

$$C_i = \begin{pmatrix} C_i(A) \\ C_i(B) \end{pmatrix} \tag{B.13}$$

Equation B.10 is a matrix eigenvalue equation ([7] of Chapter 2), where the role of the Hamiltonian operator is played by the matrix H. From Equation B.10, we can obtain $\big[H - E_i(k)S\big]C_i = 0$ which can be satisfied by $C_i = 0$ (i.e. there is no wave function) or

$$\det\big[H - E_i(k)S\big] = 0 \tag{B.14}$$

which provides the solutions for $E_i(k)$ that we seek. We must therefore evaluate the various matrix elements in H and S.

Substituting the expression in Equation B.2 into B.6 gives, for S_{AA},

$$S_{AA} = \frac{1}{N} \int \left[\sum_{l=1}^{N} e^{-ik \cdot R(A)_l} \Psi^*(A)_l \right]\left[\sum_{l=1}^{N} e^{ik \cdot R(A)_l} \Psi(A)_l \right] d\tau \tag{B.15}$$

To avoid evaluating expressions with N^2 terms, we make the reasonable simplifying assumption that each atomic wave function $\Psi(A)_l$ overlaps significantly only with those of its nearest neighbour atoms. So the wave function for each A atom overlaps with those for the three adjacent B atoms, but not with those of the nearest adjacent A atoms. Using this assumption, when we multiply out the brackets in Equation B.15 the only terms which will not integrate to zero are those where we multiply the wave functions for the same atom. Hence,

$$S_{AA} = \frac{1}{N}\sum_{l=1}^{N}\int \Psi^*(A)_l \Psi(A)_l d\tau = 1 \tag{B.16}$$

Similarly, $S_{BB} = 1$.

To evaluate H_{AA}, we begin by applying the Hamiltonian operator to $\Psi(A)_l$. We expect the $\Psi(A)_l$ wave function to be an eigenfunction of the Hamiltonian operator, with an eigenvalue of ε_{2p} (the orbital energy of the 2p level of the carbon atom in the graphene crystal). So,

$$\hat{H}\Psi(A)_l = \varepsilon_{2p}\Psi(A)_l$$

$$H_{AA} = \varepsilon_{2p} \tag{B.17}$$

Similarly, $H_{BB} = \varepsilon_{2p}$.

We must now evaluate S_{AB} and H_{AB}:

$$S_{AB} = \frac{1}{N}\int \left[\sum_{l=1}^{N}e^{-ik\cdot R(A)_l}\Psi^*(A)_l\right]\left[\sum_{l=1}^{N}e^{ik\cdot R(B)_l}\Psi(B)_l\right]d\tau \tag{B.18}$$

Here also, we can avoid an expression with N^2 terms, but we must account for the overlap of the wave functions for each A atom with its three nearest neighbour B atoms. Since every A atom is identical and every B atom is identical, we expect all N terms (one for each A atom) to have the same value and can simplify matters by considering just one A atom, that located at the origin, and the three adjacent B atoms. We can thus write

$$S_{AB} = \int \Psi^*(A)\sum_{n=1}^{3}e^{ik\cdot R_n}\Psi(B)_n d\tau \tag{B.19}$$

Here, the vectors R_n are those specified in Equation 3.7 to describe the coordinates of the three B atoms adjacent to an A atom at the origin. Hence we can write

$$S_{AB} = \int \Psi^*(A)\Psi(B)d\tau \times f(k) = sf(k) \tag{B.20}$$

where $f(k)$ is also as defined in Chapter 3 (Equation 3.9). s is the overlap integral between the wave functions for the $2p_z$ orbitals in the adjacent atoms and can be evaluated numerically based on the properties of the wave functions. Using the same methodology, we can write

$$H_{AB} = tf(k) \tag{B.21}$$

where t is also an overlap integral which can be evaluated numerically based on the properties of the wave functions. In summary,

$$s = \int \Psi^*(A)\Psi(B)d\tau \tag{B.22}$$

$$t = \int \Psi^* (A) \hat{H} \Psi (B) d\tau \tag{B.23}$$

We can then evaluate S_{BA} and H_{BA} using the Hermitian conjugation relation ([7] of Chapter 2) $H_{BA} = H_{AB}^*$, etc.:

$$S_{BA} = sf^* (k) \tag{B.24}$$

$$H_{BA} = tf^* (k) \tag{B.25}$$

Evaluating Equation B.14 using these values, we obtain

$$\begin{vmatrix} \varepsilon_{2p} - E_i(k) & tf(k) - E_i(k)sf(k) \\ tf^*(k) - E_i(k)sf^*(k) & \varepsilon_{2p} - E_i(k) \end{vmatrix} = 0$$

which we can evaluate to obtain

$$E_i(k) = \frac{\varepsilon_{2p} \pm tw(k)}{1 \pm sw(k)} \tag{B.26}$$

which is the same well-known and studied electron dispersion relation for graphene that we obtained less rigorously in Chapter 3 (Equation 3.10; Figures 1.2 and 3.2), using the same definition of $w(k)$ as in Chapter 3 $\left(w(k) = |f(k)| \right)$. The i subscript corresponds to the two different Bloch functions defined in Equation B.3, with differing energies as a result of the $\pm$ term appearing in Equation B.26.

B.3   Demonstration of Dirac-like Hamiltonian matrix for carriers near the K points in graphene

In Chapter 4, we learnt about the observation of an anomalous integer quantum Hall effect in graphene. We explained this using the expression (Equation 4.20) for the energy levels of a massless relativistic spin-1/2 particle in a magnetic field. These energy levels are the eigenvalues obtained when the Dirac equation (the relativistic equivalent of the Schrödinger equation) is solved for a particle in a magnetic field.

The reason why we cannot use the conventional treatment of the quantum Hall effect (Section 4.1.2) is because this treatment assumes that the electrons have mass. We have already seen (Chapter 3) that in fact the dispersion relation for electrons in graphene near to the Fermi level and K point is that for a massless particle.

In this section, we will learn in more detail why this dispersion relation leads us to use the Dirac equation to obtain the energy levels for the electrons in large magnetic fields (Equation 4.21). By making approximations to the tight-binding Hamiltonian matrix (Equation B.10) valid close to the K point, we will obtain a Hamiltonian matrix mathematically identical to the Hamiltonian matrix in the Dirac equation.

We start by writing down vectors leading to the K points. Whilst there are six K points at the edge of the first Brillouin zone in total (Figure B.1) most are equivalent as they can

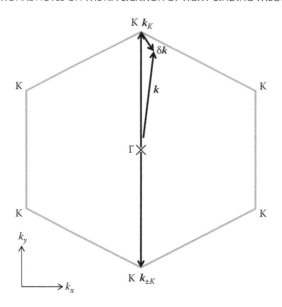

FIGURE B.1
First Brillouin zone of graphene with vectors marked leading to two inequivalent K points (defined by Equation B.27) and also the vector **k** (Equation B.29) leading to a location close to one of the K points.

be joined by the reciprocal lattice vectors (Equations 1.2 and 8.15). We therefore need to consider only two inequivalent K points. For mathematical convenience, we will choose those which lie along the k_y axis from the Γ point, at locations $k_{\pm K}$:

$$k_{\pm K} = \pm \begin{pmatrix} 0 \\ \dfrac{4\pi}{3a} \end{pmatrix} \tag{B.27}$$

From inspection of our expression for the energy (Equation B.26) and for $f(k)$ (Equation 3.9), we can draw some initial conclusions about the Hamiltonian matrix at the K points. At the K points $f(k) = 0$ and $w(k) = 0$ with $\varepsilon_{2p} = 0$ (see section 3.3). This is what causes the conduction and valence bands to meet and also causes (according to Equations B.20 through B.24)

$$H_{AB} = H_{BA} = S_{AB} = S_{BA} = 0 \tag{B.28}$$

at the K points. Physically, this means that there is no coupling between the two (real-space) sublattices made up of the A and B atoms, respectively. We can now produce an approximate Hamiltonian matrix valid close to the K point by evaluating $f(k)$ for wavevectors close to the K point. We can begin by writing down an expression for such a wavevector (see also Figure B.1):

$$k = k_{\pm K} + \delta k = k_{\pm K} + \begin{pmatrix} \delta k_x \\ \delta k_y \end{pmatrix} \tag{B.29}$$

Substituting this into our expression for $f(k)$ (Equation 3.9) and performing a first-order binomial expansion ($e^x \approx 1 + x$) on the exponential terms, we obtain

$$f(k) \approx \left(1 + \frac{i\delta k_x a}{\sqrt{3}}\right) + 2\left(1 - \frac{i\delta k_x a}{2\sqrt{3}}\right)\cos\left(\pm\frac{2\pi}{3} + \frac{a\delta k_y}{2}\right)$$

We can also approximate the cosine term* and discard the term proportional to $\delta k_x \delta k_y$ (it will be extremely small compared to the other terms) to get

$$f(k) \approx \frac{a\sqrt{3}}{2}\left(i\delta k_x \pm \delta k_y\right) \tag{B.30}$$

Recalling that $H_{AB} = tf(k)$, $H_{BA} = tf^*(k)$, and our freedom to define $\varepsilon_{2p} = 0$, we can now write down an expression for the Hamiltonian matrix (Equation B.11, etc.) close to the K point:

$$H = \frac{at\sqrt{3}}{2}\begin{pmatrix} 0 & i\delta k_x \pm \delta k_y \\ -i\delta k_x \pm \delta k_y & 0 \end{pmatrix}$$

For the purposes of comparison to the Dirac equation, we now define the momentum of the electron with wavevector k relative to that at the K point:

$$p = \begin{pmatrix} p_x \\ p_y \end{pmatrix} = \hbar \begin{pmatrix} \delta k_x \\ \delta k_y \end{pmatrix}$$

Hence,

$$H = \frac{at\sqrt{3}}{2\hbar}\begin{pmatrix} 0 & ip_x \pm p_y \\ -ip_x \pm p_y & 0 \end{pmatrix} = v_F \begin{pmatrix} 0 & ip_x \pm p_y \\ -ip_x \pm p_y & 0 \end{pmatrix} \tag{B.31}$$

where we use the equation for the group velocity v_F of electrons at the Fermi level from Equation 3.15. To continue, we now perform a binomial expansion of our equation for the energy (Equation B.26) close to the K point (as discussed in Section 3.3), with $\varepsilon_{2p} = 0$. Since we established earlier that $w(k) \to 0$ as $k \to k_{\pm K}$, this is an expansion for small $w(k)$:

$$E(k) = \pm \frac{tw(k)}{1 \pm sw(k)} \approx \pm tw(k)\left(1 \mp sw(k)\right) \approx \pm tw(k) \tag{B.32}$$

As we can see from this equation, the effect of the $sw(k)$ term on the actual energy eigenvalues is second order and can be ignored for small $w(k)$. Hence, close to the K point we can make the approximation that $S_{AB} = S_{BA} \approx 0$ and that the overlap matrix S (Equation B.12) is a unit matrix. Hence the eigenvalue equation (B.10) becomes

$$v_F \begin{pmatrix} 0 & ip_x \pm p_y \\ -ip_x \pm p_y & 0 \end{pmatrix} C_i = E_i(k)C_i \tag{B.33}$$

where, to recap,

$$C_i = \begin{pmatrix} C_i(A) \\ C_i(B) \end{pmatrix} \tag{B.13}$$

$$\Psi_i(k,r) = \sum_{A,B} C_i(A,B)\varphi_{A,B}(k,r) \tag{B.3}$$

---

* $\cos(\theta + \delta\theta) = \cos\theta + \delta\theta \dfrac{d}{d\theta}\cos\theta = \cos\theta - \delta\theta\sin\theta$

Equation B.33 is an eigenvalue equation which is mathematically identical to the Dirac equation for the energy levels of a relativistic spin-1/2 fermion. The interested reader is referred to relevant literature ([7,8] of Chapter 3) to learn about the Dirac equation – such a discussion would be beyond the scope of this text.

However, what we must note here is that whilst Equation B.33 is mathematically identical to the Dirac equation what it represents physically is quite different. The differences are as follows:

Firstly, in the actual Dirac equation, the speed of light in vacuum c appears where v_F appears in Equation B.33. Note that, whilst v_F for the electrons in graphene happens to be a very high velocity, this is not the reason we are referring to the Dirac equation. We compare (B.33) to the Dirac equation because our tight-binding treatment of the electronic dispersion relation in graphene has given us (in the vicinity of the K point) a Hamiltonian matrix for the system which happens to have the same form as the Dirac Hamiltonian. We would still treat the system in this way if t was orders of magnitude smaller and the group velocity v_F was 1 ms^{-1} instead of 1×10^6 ms^{-1}. So, it is a coincidence that the electrons in graphene move at a very high group velocity and the fact that we use the relativistic Dirac equation to describe them.

Secondly, in the actual Dirac equation, the Hamiltonian operator operates on a wave function with four components. These represent the two possible orientations of the particle's spin (Dirac found the existence of spin necessary to reconcile quantum mechanics with special relativity), and the possibility of the particle being a particle or an antiparticle. This was also found mathematically necessary by Dirac, and preceded the experimental observation of antimatter. Equations B.3, B.10 - B.13, and B.33 also involve four wave functions: The subscript i in the coefficients defines whether we are looking at the wave functions for the conduction band (i.e. electrons) or the valence band (i.e. holes) (instead of particles or antiparticles), whilst the $\pm$ sign in the Hamiltonian (B.33) dictates which K point we are observing (instead of the electron spin). This quantity is therefore referred to as "pseudospin". It has no relation to the actual spin of the electrons in graphene; this is a parameter which does not appear in the eigenvalue equation.

Reference

1. E Kaxiras, *Atomic and Electronic Structure of Solids*, CUP, Cambridge (2003).

Appendix C: Fourier Transform Treatment of Diffraction, Scherrer Broadening

C.1 Fourier transform treatment of diffraction

It is relatively straightforward to perform a general treatment of diffraction from a crystal which shows that the reciprocal space lattice is the Fourier transform of the real-space lattice. Once this feature is understood, most features of diffraction experiments become self-evident. As a reference point, let us recall the relationship between a function $f(v)$ and its Fourier transform $F(w)$:

$$F(w) = \frac{1}{\sqrt{2\pi}} \int f(v) e^{iv \cdot w} d\tau \qquad (C.1)$$

$$f(v) = \frac{1}{\sqrt{2\pi}} \int F(w) e^{-iv \cdot w} d\tau$$

Here, we can of course choose how the $(1/2\pi)$ factor is split between the two transforms and on which exponent we place the minus sign. We begin by considering the diffraction problem from Figure 8.2 in more general terms. Instead of considering discrete atoms, let us consider the sample as a function $\rho(r)$ representing the amplitude of scattering of the incident radiation at each location in the sample. So $\rho(r)d\tau$ is the amplitude of scattering from the part of the sample within the volume element $d\tau$ located at r.

$\rho(r)$ depends on what probe we are diffracting from the sample. X-rays are scattered entirely by the electrons in the sample, so for an X-ray diffraction experiment $\rho(r)$ would be the distribution of electronic charge throughout the sample. We could obtain this from the electron wave functions. In a neutron diffraction experiment on the other hand, $\rho(r)$ would be the density of nuclear matter because neutrons interact with the sample only via the strong nuclear force. For electron diffraction, $\rho(r)$ would be some combination of the nuclear charge distribution and the electronic charge distribution for the sample as electrons interact with both the nuclei and the electrons in the sample.

Let us consider (Figure C.1) the amplitude dE of the diffracted beam at a point R from the volume element $d\tau$. Analogously to Equation 8.2 we can write

$$dE = E_0 e^{i(k \cdot r - \omega t)} \times \rho(r) d\tau \times \frac{e^{ik|R-r|}}{|R-r|} \qquad (C.2)$$

We can proceed in a manner analogous to the procedure in Chapter 8 to obtain

$$dE = \rho(r) d\tau e^{-iK \cdot r} \qquad (C.3)$$

where we have again ignored the factors that are the same for all locations in the crystal and defined the scattering vector K:

$$K = k'_r - k$$

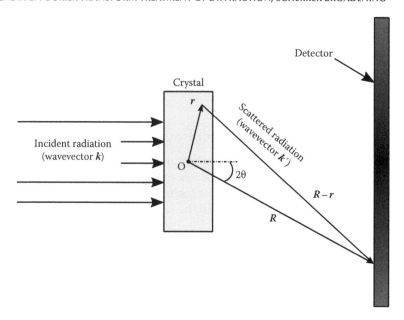

FIGURE C.1
Experimental geometry for a diffraction experiment.

where now we use the notation k'_r to denote the wavevector of the beam scattered from location r. The expression in Equation C.3 can be integrated to obtain the total diffracted amplitude for a given R (i.e. for a given K):

$$E(K) = \int \rho(r) e^{-iK \cdot r} d\tau \tag{C.4}$$

Comparing this expression to Equation C.1, we can see that $E(K)$, the function representing the diffracted intensity and therefore the reciprocal space lattice, is simply the Fourier transform of the real-space lattice as represented by $\rho(r)$.

In Chapter 8, we considered the scattering on an atom-by-atom basis using the atomic form factor. We will now see how this is consistent with the Fourier transform approach mentioned earlier. Using the atomic form factor approach, we can approximate $\rho(r)$ with a δ-function at the location of each atom, multiplied by the atomic form factor to represent the nature of the scattering from each atom. Hence

$$\rho(r) = \sum_n \delta(r - r_n) f_n \tag{C.5}$$

Substituting Equation C.5 into C.4 and recalling [1] that $\int \delta(r - r_n) d\tau = 1$, we obtain

$$E(K) = \sum_n f_n e^{-iK \cdot r_n} \tag{C.6}$$

This is identical to Equation 8.4, from which we can proceed to obtain the Laue conditions for diffraction.

C.2   Scherrer broadening

Textbook treatments of diffraction such as the Bragg treatment, and the Laue treatment that we performed in Chapter 8, assume that the real-space lattice is infinitely large. They predict diffraction peaks which are essentially δ-functions, with a diffracted beam only present at the specific values of θ or K that satisfy the diffraction conditions. But any real diffraction peak has a finite width for a variety of reasons such as the incoming beam not being perfectly collimated and monochromatic, and imperfections in the lattice due to defects or thermal motion of the atoms. There is one mechanism for broadening of diffraction peaks which is of particular relevance to the study of nanomaterials and that is Scherrer broadening [2], broadening due to the fact that the real-space lattice is not infinitely large.

The physical reasons why Scherrer broadening occurs, and the reason why the severity of the broadening is related to the wavelength of the incoming radiation, can be understood simply by referring to the Bragg construction. In the Bragg construction (Figure C.2), a diffraction peak is present when the path difference ABC for reflection from the successive layers of atoms in the sample ($2d \sin \theta$) is some multiple of the wavelength of the incoming beam ($n\lambda$), so the beams reflected off successive layers of atoms interfere constructively.

But what happens when the Bragg condition is nearly met, when the beams reflected from successive layers are (for instance) 1° out of phase? In this case, we expect the beam reflected off the 180th layer to interfere destructively with the beam reflected from the first layer, thus preventing the observation of a diffraction peak. When there are only 10 atomic layers present, however, the beams reflected from successive layers will be at most 10° out of phase, so an intense diffracted beam will still be observed. This is the phenomenon of Scherrer broadening (Figure C.3). In this figure, we see extremely narrow X-ray diffraction peaks from a bulk sample of CdSe and much broader X-ray diffraction peaks from CdSe nanoparticles.

Scherrer broadening is less severe if the incident radiation has a shorter wavelength. We can also understand this using the Bragg construction. Suppose that, for a given scattering angle the Bragg condition is satisfied (i.e. $2d \sin \theta = n\lambda$). Now imagine that we look at a slightly different scattering angle. The change in scattering angle will lead to a certain optical path difference between the routes taken by the beam diffracted from successive

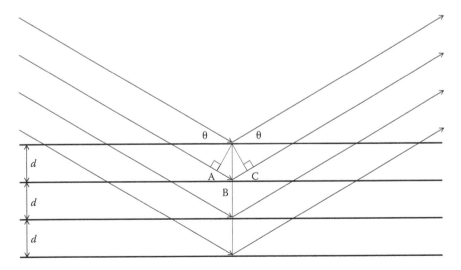

FIGURE C.2
Bragg construction for diffraction.

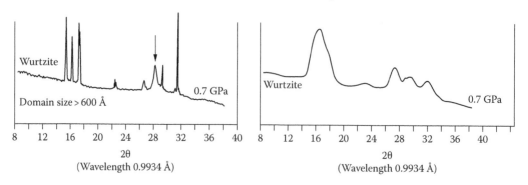

FIGURE C.3

Example of Scherrer broadening in X-ray diffraction peaks. (a) X-ray diffraction from bulk CdSe in the wurtzite phase. (b) X-ray diffraction from CdSe nanocrystals in the wurtzite phase demonstrating the effect of Scherrer broadening on the peak width. (Reprinted with permission from Tolbert, S.H. and Alivisatos, A.P., *J. Chem. Phys.*, 102, 4642. Copyright 1995, American Institute of Physics.)

layers, depending on the change in scattering angle and the inter-layer spacing *d*. If we use radiation with a shorter wavelength, the path difference will lead to a larger phase difference between radiation scattered from successive layers. Hence destructive interference eliminates the diffracted beam for a smaller change in scattering angle θ.

A formal mathematical treatment of the aforementioned points [2] leads to an expression for the diffraction peak width due to Scherrer broadening:

$$\Delta(2\theta) = \frac{\lambda}{Nd\cos\theta_0}$$

where

λ is the wavelength of the incident radiation
N is the number of atomic layers in the sample
d is the inter-layer spacing
θ_0 is the scattering angle for which the Bragg condition is satisfied
$\Delta(2\theta)$ is the peak width in units of 2θ due to Scherrer broadening

References

1. ML Boas, *Mathematical Methods in the Physical Sciences*, Wiley, Hoboken (1966).

2. A Guinier, *X-Ray Diffraction in Crystals, Imperfect Crystals, and Amorphous Bodies*, Dover, Mineola (1964).

3. SH Tolbert and AP Alivisatos, *J. Chem. Phys.* **102**, 4642 (1995).

Bibliography

Textbooks on quantum mechanics and solid state physics

PW Atkins, *Physical Chemistry*, OUP (1982).

An undergraduate level textbook with a good introduction to molecular orbital theory.

VB Berestetsky, EM Lifshitz and LP Pitaevsky, *Relativistic Quantum Theory*, Pergamon Press (1971).

English translation of the classic textbook on relativistic quantum theory; volume IV of the course on theoretical physics by Landau and Lifshitz, based on lecture courses given at Moscow State University in the 1959–1960 academic year. Calculation of the Landau energy levels for a relativistic fermion in a magnetic field is an example given in the textbook, and which is now used to understand the anomalous quantum Hall effect in graphene.

JR Hook and HE Hall, *Solid State Physics*, Wiley-VCH (1998).

Undergraduate level textbook with good coverage of lattice vibrations, diffraction and the quantum Hall effect (though not in graphene).

E Kaxiras, *Atomic and Electronic Structure of Solids*, CUP (2003).

Graduate level textbook with extensive coverage of tight-binding theory, electronic properties of solids in general and lattice vibrations.

L Lyons, *All You Wanted to Know about Mathematics but Were Afraid to Ask (Mathematics for Science Students)*, CUP (1998).

A very readable undergraduate level textbook in two volumes detailing the useful mathematics for physical science students to learn. Written by a physics lecturer, the text has very good introductions to topics such as matrices, Fourier transforms and vector operators.

T Ohlsson, *Relativistic Quantum Physics*, CUP (2011).

A recent textbook on relativistic quantum physics, with good coverage of the Dirac equation.

AIM Rae, *Quantum Mechanics*, IoP Publishing (1992).

An undergraduate level textbook on quantum mechanics. Good coverage of the solution of the Schrödinger equation for the hydrogenic atom, and the origin of the spherical harmonic functions that determine the shape of the electron wavefunctions for orbitals such as the 2s, $2p_x$ etc.

HM Rosenberg, *The Solid State*, OUP (1975).

Introductory solid state physics textbook concentrating more on key physical concepts than mathematical derivation. Particularly good section in propagation of waves in crystals outlining the parallels between diffraction, and propagation of phonons and electrons in a crystal.

J Singleton, *Band Theory and Electronic Properties of Solids*, OUP (2001).

Undergraduate level textbook with extensive coverage of cyclotron resonance and the quantum Hall effect (though not in graphene), and a good introduction to the density of states concept. The text has a lot of general information about electronic dispersion relations (band structure) and a short introduction to tight binding methods.

PY Yu and M Cardona, *Fundamentals of Semiconductors*, Springer (1996).

Classic textbook on semiconductor physics including electronic dispersion relations, vibrational properties, Raman and Brillouin scattering, the quantum Hall effect and many other important topics. Written at a good level for research students and final year undergraduates.

Textbooks on graphene and carbon nanotubes

MI Katsnelson, *Graphene: Carbon in Two Dimensions*, CUP (2012).

Standard graduate level textbook, concentrating particularly on the electronic properties of graphene.

S Reich, C Thomsen and J Maultzsch, *Carbon Nanotubes: Basic Concepts and Physical Properties*, Wiley-VCH (2004).

Graduate level textbook covering knowledge (as of 2004) of electronic and vibrational properties of SWCNTs. Has some coverage of photoluminescence from individual SWCNTs, but predates more recent developments in our ability to apply this technique in the (n,m) characterization of SWCNTs. Synthesis of SWCNTs is not covered.

R Saito, G Dresselhaus and MS Dresselhaus, *Physical Properties of Carbon Nanotubes*, ICP (1998).

Graduate level textbook with good coverage of orbital hybridization in carbon, electronic dispersion relations of SWCNTs (based on application of tight binding theory to graphene), phonons and Raman scattering in SWCNTs. The book does, however, predate important developments in our understanding of SWCNTs and experimental abilities in this regard. In particular, it predates the observation of strong photoluminescence from individual SWCNTs, the knowledge that the energies of these optical transitions are reduced by the existence of excitons, our ability to separate SWCNTs from the bundle and sort them by diameter, and to control (n,m) during growth.

Textbook chapters and review publications

Scientific Background on the Nobel Prize in Physics 2010, Royal Swedish Academy of Sciences (2010).

Citation for the award of the Nobel Prize in Physics 2010 to KS Novoselov and AK Geim for their ground-breaking experiments since 2004 regarding the discovery and characterization of graphene.

PT Araujo, PBC Pesce, MS Dresselhaus, K Sato, R Saito and A Jorio, Resonance Raman spectroscopy of the radial breathing modes in carbon nanotubes, *Physica E* **42**, 1251 (2010).

Detailed review of our knowledge of the radial breathing mode in the Raman spectrum of the SWCNT, focussing on its relation to tube diameter.

P Avouris, M Freitag and V Perebeinos, Carbon-nanotube photonics and optoelectronics, *Nat. Photonics* **2**, 341 (2008).

Review paper covering the optical properties of SWCNTs and their potential applications in photonics and optoelectronics.

AH Castro Neto, F Guinea, NMR Peres, KS Novoselov and AK Geim, The electronic properties of graphene, *Rev. Mod. Phys.* **81**, 109 (2009).

A comprehensive review paper on the electronic properties of graphene, also including a detailed discussion of the role of out-of-plane phonons in preventing free-standing graphene from being stable.

RS Edwards and KS Coleman, Graphene synthesis: Relationship to applications, *Nanoscale* **5**, 38 (2013).

Review of methods for graphene synthesis and exfoliation.

AC Ferrari and DM Basko, Raman spectroscopy as a versatile tool for studying the properties of graphene, *Nat. Nanotechnol.* **8**, 235 (2013).

Comprehensive and readable review paper covering Raman spectroscopy of graphene, from the basics to the current state of the art.

AK Geim and KS Novoselov, The rise of graphene, *Nat. Mater.* **6**, 183 (2007).

Readable review paper covering the background to graphene research (why 2D materials were presumed not to exist) and the 3 years of graphene research from 2004 to 2007, with emphasis on experimental advances leading to isolation of graphene and the materials' electronic properties.

SA Hodge, MK Bayazit, KS Coleman and MSP Shaffer, Unweaving the rainbow: A review of the relationship between single-walled carbon nanotube molecular structures and their chemical reactivity, *Chem. Soc. Rev.* **41**, 4409 (2012).

Tutorial review on the relationship between SWCNT (n,m) indices and chemical reactivity, a relationship which allows possible separation of SWCNTs according to (n,m) after synthesis.

F Karlický, KKR Datta, M Otyepka and R Zbořil, Halogenated graphenes: Rapidly growing family of graphene derivatives, *ACS Nano* **7**, 6434 (2013).

A review paper on the synthesis and properties of halogenated graphenes.

H Kroto, Carbyne and other myths about carbon, *Chem. World* (November 2010).

Article detailing the history of the various claims (since 1967) that carbon has been isolated in sp-bonded form (i.e. carbyne), along with the refutation of these claims.

E McCann, Chapter 8: Electronic properties of monolayer and bilayer graphene, in *Graphene Nanoelectronics*, H Raza (ed.), Springer (2012).

Graduate level textbook chapter applying tight-binding theory to graphene and justifying use of the Dirac equation in describing the electronic dispersion relation near the Fermi level.

M Meyyappan, A review of plasma enhanced chemical vapour deposition of carbon nanotubes, *J. Phys. D: Appl. Phys.* **42**, 213001 (2009).

2009 review paper on plasma-enhanced CVD of carbon nanotubes.

M Monthioux and VL Kuznetsov, Who should be given the credit for the discovery of carbon nanotubes?, *Carbon* **44**, 1621 (2006).

A review article detailing the history of carbon nanotube research prior to 1991.

V Nicolosi, M Chhowalla, MG Kanatzidis, MS Strano and JN Coleman, Liquid exfoliation of layered materials, *Science* **340**, 1226419 (2013).

Review of the liquid phase exfoliation of graphene and other layered materials.

KS Novoselov, VI Fal'ko, L Colombo, PR Gellert, MG Schwab and K Kim, A roadmap for graphene, *Nature* **490**, 192 (2012).

Overview of progress in graphene research to 2012 and critical analysis of a range of potential applications for graphene.

L-M Peng, Z Zhang and S Wang, Carbon nanotube electronics: Recent advances, *Mater. Today* **17**, 433 (2014); GS Tulevski et al., Toward high-performance digital logic technology with carbon nanotubes, *ACS Nano* **8**, 8730 (2014); MFL De Volder et al., Carbon nanotubes: Present and future commercial applications, *Science* **339**, 535 (2013).

A selection of recent review papers discussing the progress of SWCNTs towards commercial applications.

M Pumera and CH An Wong, Graphane and hydrogenated graphene, *Chem. Soc. Rev.* **42**, 5987 (2013).

A tutorial review on the hydrogenation of graphene-related materials, though with an emphasis on hydrogenation of few-layer graphene and graphite.

JM Tour, Seeds of selective nanotube growth, *Nature* **512**, 30 (2014); H Wang et al., Catalysts for chirality selective synthesis of single-walled carbon nanotubes, *Carbon* **81**, 1 (2015); F Zhang, P-X Hou, C Liu and H-M Cheng, Epitaxial growth of single-wall carbon nanotubes, *Carbon* **102**, 181 (2016).

A selection of publications commenting on recent progress in chirality-selective synthesis of SWCNTs.

RJ Young, IA Kinloch, L Gong, KS Novoselov, The mechanics of graphene nanocomposites: A review, *Compos. Sci. Technol.* **72**, 1459 (2012).

A review article describing current technology on the incorporation of graphene into nanocomposite materials and the obstacles to achieving good stress transfer from the host material to the graphene. The preparation methods, and mechanical properties of both graphene and graphene oxide are reviewed in this context.

Original research publications

VI Artyukhov, ES Penev and BI Yakobson, Why nanotubes grow chiral, *Nat. Commun.* **5**, 4892 (2014).

Paper proposing theoretical explanation for a long-standing puzzle: Why is there a preference towards SWCNTs forming with near-armchair chiral indices?

P Blake et al., Making graphene visible, *Appl. Phys. Lett.* **91**, 063124 (2007).

Study characterizing in detail the weak interference effect between graphene layers and a thin SiO_2 substrate which is responsible for making mono-layer graphene visible in an optical microscope.

DC Elias et al., Control of graphene's properties by reversible hydrogenation: Evidence for graphane, *Science* **323**, 610 (2009).

Publication describing the reversible chemical modification of pristine graphene and conversion to sp^3 bonding by addition of hydrogen. See also *Beyond the Wonder Material* KS Novoselov, Physics World (August 2009).

JA Elliott, JKW Sandler, AH Windle, RJ Young and MSP Shaffer, Collapse of single-wall carbon nanotubes is diameter dependent, *Phys. Rev. Lett.* **92**, 095501 (2004).

A combined theoretical and experimental work detailing the instability of large (>5 nm) SWCNTs against collapse. The work demonstrates that the SWCNT structure is only stable on an extremely small scale.

AC Ferrari et al., Raman spectrum of graphene and graphene layers, *Phys. Rev. Lett.* **97**, 187401 (2006).

Publication showing the Raman spectra of mono-layer, bi-layer and few-layer graphene samples with thickness directly calibrated using electron microscopy and diffraction. The clear evolution of the characteristics of the spectra with the number of layers is demonstrated and explained in terms of the double resonant Raman scattering process.

M Hanfland, H Beister and K Syassen, Graphite under pressure: Equation of state and first order Raman modes, *Phys. Rev. B* **39**, 12598 (1989).

Paper measuring elastic moduli and Grüneisen parameters of graphite by performing X-ray diffraction and Raman spectroscopy on graphite in a diamond anvil high pressure cell.

Y Hernandez et al., High-yield production of graphene by liquid-phase exfoliation of graphite, *Nat. Nanotechnol.* **3**, 563 (2008).

The production of graphene using liquid-phase exfoliation is reported and the required solvent properties (particularly in terms of surface energy) are discussed in detail.

AG Kvashnin, LA Chernatonskii, BI Yakobson and PB Sorokin, Phase diagram of quasi-two-dimensional carbon, from graphene to diamond, *Nano Lett.* **14**, 676 (2014).

Publication presenting an interesting alternative view on the hydrogenation of graphene: The transition from sp^2 bonding to sp^3 bonding upon hydrogenation of graphene is related to the transition from sp^2 bonding to sp^3 bonding when bulk graphite is transformed into diamond at high pressure and temperature.

C Lee, X Wei, JW Kysar and J Hone, Measurement of the elastic properties and intrinsic strength of monolayer graphene, *Science* **321**, 385 (2008).

A direct measurement of the strength of mono-layer graphene. The measurement was achieved through stretching a suspended graphene layer to breaking point by exerting force on it using an atomic force microscope tip.

X Li et al., Large-area synthesis of high-quality and uniform graphene films on copper foils, *Science* **324**, 1312 (2009).

Report of the large scale synthesis of uniformly mono-layer graphene films on copper foil by CVD.

K Liu et al., An atlas of carbon nanotube optical transitions, *Nat. Nanotechnol.* **7**, 325 (2012).

A comprehensive work experimentally measuring the different optical transition energies in SWCNTs with over 200 different (*n,m*) assignments.

MM Lucchese et al., Quantifying ion-induced defects and Raman relaxation length in graphene, *Carbon* **48**, 1592 (2010).

Detailed experimental study on the evolution of the Raman spectra of graphene as a function of the concentration of defects.

JC Meyer et al., On the roughness of single- and bi-layer graphene membranes, *Solid State Commun.* **143**, 101 (2007).

Publication describing detailed findings of electron diffraction and transmission electron microscopy experiments on mono-layer and bi-layer graphene. The ability to identify mono-layer and bi-layer graphene samples from diffraction data is shown and the evidence provided by diffraction data for the intrinsic rippling of graphene sheets is also presented.

TMG Mohiuddin et al., Uniaxial strain in graphene by Raman spectroscopy: G peak splitting, Grüneisen parameters, and sample orientation, *Phys. Rev. B* **79**, 205433 (2009).

A detailed and careful combined experimental and theoretical study elucidating the Grüneisen parameters for graphene.

N Mounet and N Marzari, High-accuracy first-principles determination of the structural, vibrational and thermodynamical properties of diamond, graphite, and derivatives, *Phys. Rev. B* **71**, 205214 (2005); D Yoon, Y-W Son and H Cheong, Negative thermal expansion coefficient of graphene measured by Raman spectroscopy, *Nano Lett.* **11**, 3227 (2011).

Theoretical and experimental works characterizing the negative thermal expansion coefficient of graphene.

RR Nair et al., Fine structure constant defines visual transparency of graphene, *Science* **320**, 1308 (2008).

In Chapter 4, we discuss how certain electronic properties of graphene are defined only by fundamental constants. The same is true of its visual transparency, which is defined only by the fine structure constant $= e^2/hc$.

KS Novoselov et al., Electric field effect in atomically thin carbon films, *Science* **306**, 666 (2004).

The authors describe the mechanical exfoliation technique, which allowed the reliable isolation of graphene mono-layers of extremely high crystalline quality and hence started the goldrush of scientific interest in graphene over the past decade. The electric field effect and cyclotron motion of electrons were also observed in graphene.

KS Novoselov et al., Two-dimensional gas of massless Dirac fermions in graphene, *Nature* **438**, 197 (2005); Y Zhang, Y-W Tan, HL Stormer and P Kim, Experimental observation of the quantum Hall effect and Berry's phase in graphene, *Nature* **438**, 201 (2005).

Publications simultaneously reporting the observation of cyclotron resonance and quantum Hall effect in graphene, strong evidence for the linearity of the electronic dispersion relation in the vicinity of the Fermi level and exceptionally high group velocity of electrons in graphene.

A Oberlin, Graphitization of thin carbon films, *Carbon* **22**, 521 (1984); J Goma and M Oberlin, Carbonization and graphitization, *Thin Solid Films* **65**, 221 (1980).

The original publications elucidating the role of heat treatment in the formation of ordered graphene layers in graphite, knowledge now applied to SWCNT and graphene synthesis.

MJ O'Connell et al., Band gap fluorescence from individual single-walled carbon nanotubes, *Science* **297**, 593 (2002).

Publication describing how the SWCNT bundle can be broken apart using ultrasound, enabling the observation of photoluminescence from individual SWCNTs.

R Saito, G Dresselhaus and MS Dresselhaus, Trigonal warping effect of carbon nanotubes, *Phys. Rev. B* **61**, 2981 (2000).

An exploration of the consequences of the quantization of electron wavevector in single-walled carbon nanotubes, in particular, its effect on the density of states. The text provides a readable introduction to this essential topic, but predates the discovery of the role of excitons in SWCNTs.

JR Sanchez-Valencia et al., Controlled synthesis of single-chirality carbon nanotubes, *Nature* **512**, 61 (2014); F Yang et al., Chirality-specific growth of single-walled carbon nanotubes on solid alloy catalysts, *Nature* **510**, 522 (2014); F Yang et al., Growing zigzag (16,0) carbon nanotubes with structure-defined catalysts, *JACS* **137**, 8688 (2015).

Three key papers from 2014 onwards reporting the long sought-after ability to control (n,m) indices during SWCNTs synthesis.

MM Shulaker et al., Carbon nanotube computer, *Nature* **501**, 526 (2013).

Report of the development of a computer utilizing SWCNT transistors. An illustration of the formidable obstacles to the commercial applications of SWCNTs in electronics; each transistor in this computer consisted of an array of SWCNT with different (n,m) in which the metallic tubes had been burnt off and was ca. 10 μm diameter. In contrast, 20 nm diameter silicon transistors are utilized in commercially available electronic devices.

F Wang, G Dukovic, LE Brus and TF Heinz, The optical resonances in carbon nanotubes arise from excitons, *Science* **308**, 838 (2005).

Publication demonstrating that the optical resonances in SWCNTs are due to the existence of excitons. This is essential knowledge to understand the exact transition energies at which SWCNTs will emit and absorb light, crucial to the use of photoluminescence and Raman spectroscopy in (n,m) assignment and to potential applications of SWCNTs in optoelectronics.

D Yoon, Y-W Son and H Cheong, Negative thermal expansion coefficient of graphene measured by Raman spectroscopy, *Nano Lett.* **11**, 3227 (2011).

Publication describing experimental measurement of the thermal expansion coefficient of graphene, including a good summary of the various experimental and theoretical issues entailed.

D Zhang et al., (n,m) Assignments and quantification for single walled carbon nanotubes on SiO_2/Si substrates by resonant Raman spectroscopy, *Nanoscale* **7**, 10719 (2015).

A study laying out in detail a procedure for assigning (n,m) indices to SWCNT using resonant Raman spectroscopy. Similar methodologies have been proposed in earlier studies dating back as far as 2001. However, this study takes full account of recent developments in our understanding of SWCNTs. In particular, the influence of the SWCNT environment on the radial breathing mode frequencies and resonant optical transition energies. See also H Telg et al., Chiral index dependence of the G^+ and G^- Raman modes in semiconducting carbon nanotubes, *ACS Nano* **6**, 904 (2012).

Index